# Induction of
# Bone Formation
# in Primates

## The Transforming Growth Factor-beta 3

# Induction of Bone Formation in Primates

## The Transforming Growth Factor-beta 3

Edited by

## Ugo Ripamonti

**CRC Press**
Taylor & Francis Group
Boca Raton London New York

CRC Press is an imprint of the
Taylor & Francis Group, an **informa** business

CRC Press
Taylor & Francis Group
6000 Broken Sound Parkway NW, Suite 300
Boca Raton, FL 33487-2742

First issued in paperback 2019

© 2016 by Taylor & Francis Group, LLC
CRC Press is an imprint of Taylor & Francis Group, an Informa business

No claim to original U.S. Government works

ISBN-13: 978-1-4822-1616-5 (hbk)
ISBN-13: 978-0-367-37740-3 (pbk)

**Visit the Taylor & Francis Web site at**
**http://www.taylorandfrancis.com**

**and the CRC Press Web site at**
**http://www.crcpress.com**

# Contents

## 3  Rapid Induction of Bone Formation by the Transforming Growth Factor-β₃ Isoform    47

UGO RIPAMONTI

## 4  Induction of Bone Formation by the Mammalian Transforming Growth Factor-βs: Molecular and Morphological Insights    75

RAQUEL DUARTE, KURT LIGHTFOOT,
AND UGO RIPAMONTI

# Foreword: A Short History of Bone and Embryonic Induction

The repair of bone fractures provides one of the best examples of tissue regeneration in adult humans. It is of great medical interest to repair bone in those cases of fractures that fail to heal or in wounds involving large bone defects. Orthopedic surgeons have been interested in the biology of bone regeneration for many decades, as explained in this timely book that relates the unexpected discovery by Ugo Ripamonti of a central role of transforming growth factor-βs (TGF-βs) in bone regeneration. Remarkably, these novel insights in the treatment of bone fractures were achieved by a research team in faraway South Africa, under the blue skies of the Witwatersrand plateau near Johannesburg. They used similarities between human biology and the baboon to make discoveries that would not have been possible in other mammalian models.

The TGF-β superfamily constitutes the largest family of growth factors in humans, with a total of 33 different secreted ligands. They are commonly divided into the bone morphogenetic protein (BMP) and the TGF-β branches. Of these, the TGF-β themselves (TGF-β1 to -3) are the most recent evolutionary acquisitions. This family of ligands is not solely of interest in orthopedic medicine, but has proven of enormous importance during the signaling that takes place in the early embryo, in which a single cell, the fertilized egg, is converted into a body plan containing many tissue types. Curiously, it was the search for agents that repair fractures that led to the isolation of many of the diffusible substances that control embryonic development.

The story told in this book starts in 1938 with the publication by Gustave Levander, from Uppsala, Sweden, of a study of bone regeneration. He showed that the regenerating callus of long bones could be cut into tiny fragments, fixed in ethanol, stored for several days, diluted to 40% with water, and injected into the thigh muscles of rabbits. He made the astonishing finding that these dead tissues were able to induce bone and cartilage from muscle mesenchyme in 22% of the cases.

In his paper, Levander made the intellectual connection between the induction of bone in mesenchyme and the studies on embryonic development by Hans Spemann of Freiburg University, who had recently received the 1935 Nobel Prize for Medicine. Spemann and his student Hilde Mangold had shown that a region of the gastrula embryo, called the organizer, was able to change the differentiation of neighboring cells into different tissue types. He called this process embryonic induction. During the 1930s, many attempts were made to isolate the mysterious inducing substances from embryos, but all met with failure. It was the work on bone that was to lead the way in the isolation of embryonic inducers. Levander concluded his paper (1938) in this way: "On the basis of these experiments the author holds that bone regeneration takes place as the result of some specific bone forming substance activating the non-specific mesenchymal tissue. The theory also agrees with the views advanced by Spemann with regard to embryonic development."

One can only stand in awe of the prescience of Levander's insight, for it was only in the 1990s that Noggin and Chordin, two BMP antagonists secreted by organizer tissue, were found to mediate Spemann's embryonic induction. Noggin (isolated by Richard Harland at the University of California–Berkeley) and Chordin (isolated by us at the University of California–Los Angeles) generate a gradient of BMP signaling that is maximal in the ventral side of the embryo. In fact, dorsal–ventral patterning has been found to be the ancestral function of BMPs in all bilateral animals. For example, in *Drosophila* the function of the morphogen Decapentaplegic (Dpp) can be replaced by human BMP-2 or BMP-4. Teleost fish co-opted this ancient dorsal–ventral signaling system to generate an evolutionarily novel tissue, bone.

The next key step in this saga was the discovery in 1965 by orthopedic surgeon Marshall Urist, of the University of California–Los Angeles, that the matrix of long bones decalcified by incubation in 0.6 N HCl over many days had very powerful bone-inducing activity. While dead bone fragments only generated a low proportion of bone induction cases after transplantation into abdominal muscle of rabbits, implanted bone matrix was effective in up to 90% of the cases. Furthermore, decalcified matrix could be used to bridge regeneration in long bone defects. He proposed the name BMP for this inducing substance, which much later was found to be a key morphogenetic protein in the embryo as well as in bone.

BMPs proved difficult to separate from the matrix until 1981, when Hari Reddi, then at the National Institutes of Health (Bethesda, Maryland), discovered that BMPs could be solubilized with 4 M guanidine–HCl and 8 M urea. The protein could now be purified biochemically, but in order to restore its bone-inducing activity, it was necessary to implant it together with BMP-depleted matrix. This assay eventually allowed Dr. Sampath at Creative Biomolecules, Hopkinton, Massachusetts, to purify osteogenic protein-1 (OP-1), also known as BMP-7. In 1988, a team from Genetics Institute headed by John Wozney was able to obtain highly purified BMP fractions that yielded protein sequences for BMP-2 to -7. This work revealed that BMPs were members of the TGF-$\beta$ superfamily, which could now be produced in large amounts by recombinant DNA technology.

The availability of BMP proteins triggered herculean efforts by the biotech industry to apply BMPs in bone and dental regeneration. Surprisingly, so far the bone-inducing effects of BMPs in humans have yielded many disappointments as practical therapeutic agents. However, BMPs continue to have great promise in the treatment of many diseases, in particular now that it is known that BMP-6 circulates in blood. One of the main problems in the application of BMPs to bone fracture repair seems to be the induction of BMP antagonists such as Noggin by high concentrations of the growth factor in the target tissues.

It was therefore a great surprise when the Ripamonti team found that TGF-$\beta$s greatly cooperated with BMP, and were even able to induce bone on their own in the baboon system. While the expectation was that BMPs would be the main bone inducers, it may well be TGF-$\beta$ that provides the therapeutic breakthrough. The BMP branch of the pathway signals by activating receptors that phosphorylate and activate transcription factors known as Smad-1/5/8. The TGF-β branch of the pathway phosphorylates Smad-2/3 was considered completely independent of BMP. In fact, in many contexts, BMP and TGF-β oppose each other. One exception was the discovery by Peter ten Dijke in Holland that a BMP receptor can phosphorylate Smad-2 in addition to Smad-1, although with slower kinetics. There are seven type-1 receptors that must match with five type-2 receptors in order to transduce the entire spectrum of TGF-β superfamily ligands, so there is room for unexpected mixing and matching of signals. In addition, there are many opportunities for sequential induction of genes in the pathway, as indicated by the recent finding that recombinant human TGF-β$_3$ upregulates endogenous BMP-2.

In sum, this book marks a contribution to the fascinating history of BMP and TGF-β research at the intersection of molecular biology, tissue induction, bone regeneration, and orthopedic surgery.

**Edward M. De Robertis**
*Pacific Palisades, California*

## Reference

Levander G., 1938. A study of bone regeneration. *Surg. Gynecol. Obstet.* 78, 705–714.

# Contributors

**Raquel Duarte**
Department of Internal Medicine
School of Clinical Medicine
Faculty of Health Sciences
University of the Witwatersrand
Johannesburg, South Africa

**Carlo Ferretti**
Bone and Research Laboratory
School of Oral Health Sciences
Division of Maxillofacial
    and Oral Surgery
Chris Hani Baragwanath Hospital
University of Witwatersrand
Johannesburg, South Africa

**Kurt Lightfoot**
Department of Internal Medicine
School of Clinical Medicine
Faculty of Health Sciences
University of the Witwatersrand
Johannesburg, South Africa

**Jean-Claude Petit**
Bone Research Laboratory
School of Oral Health Sciences
Faculty of Health Sciences
University of the Witwatersrand
Johannesburg, South Africa

**Ugo Ripamonti**
Bone Research Laboratory
School of Oral Health Sciences
Faculty of Health Sciences
University of the Witwatersrand
Johannesburg, South Africa

# CHAPTER ONE

# Introduction

## The Discovery of Redundancy and the Bone Induction Prerogative of the Three Mammalian Transforming Growth Factor-β Isoforms

*Ugo Ripamonti*

Bone Research Laboratory, School of Oral Health
Sciences, Faculty of Health Sciences, University of the
Witwatersrand, Johannesburg, Parktown, South Africa

### 1.1 Induction of Bone Formation before Redundancy and the Discovery of the Induction of Bone Formation by the Mammalian Transforming Growth Factor-β Isoforms

Almost 27 years have elapsed since working under the *aegis* of the University of the Witwatersrand, Johannesburg, at the Dental Research Institute as a young and growing scientist from Milano University, where I qualified in medicine and surgery, odontostomatology, and maxillofacial surgery. I then reached the pleasant but very complicated and violent shores of the South African continent, where racial battles—unknown to me as a Latin descendant—were ravaging the country and its universities.

Though the historical movements were rapidly unfolding, the university still maintained a high level of organizational structure and discipline that for me were essential to understand the global power of the Anglo-Saxon culture and instrumental in learning, although with difficulties, the British language, an instrument of knowledge and culture I wanted to master to finally become a citizen of the world and grow out of the cocoon I created in Milano, safe with my family and my prospective work in the well-organized private dental practice of Papa' Cesare.

My transfer and relocation to South Africa was to be for three years, until I obtained the master of dentistry in oral medicine and periodontology, after which I planned to return to Milano to run papa's private practice. Plans change, however, as often happens in life, and in February 1983, after meeting Tracey, my willingness to return to Italy somehow vanished—because I fell in love not only with Tracey, but also with the marvelous, clear, transparent, blue skies of the beautiful then Transvaal. So I decided to stay, but I did not have a salaried position. I did, however, have the unique opportunity to be awarded a foreign visiting scientist position offered by Prof. P.E. Cleaton-Jones, former director of the Dental Research Institute and first mentor of my South African scientific adventure.

I then started to read scientific papers and browse what I thought to be important scientific contributions, trying to pick up a valid doctoral project. When composing a manuscript for the *Journal of Periodontal Research* on autotransplanted roots coated with fibrin–fibronectin after surface demineralization, which was part of my master of dentistry, I came across the classic paper of Hari Reddi and Charles Huggins published in the *Proceedings of the National Academy of Sciences USA* in 1972, a paper that significantly referred to a paper by Marshall R. Urist published in *Science* in 1965: "Bone: Formation by Autoinduction."

So, as it was by serendipity, I was suddenly exposed to the fascinating scenario of Urist's bone formation by autoinduction, and my approach to the scientific problem was that of a clinician since I was not a scientist by training, but a medical doctor with three specialties in the craniofacial area, with a keen interest in hominid evolution, powerful fast-riding motorbikes, and a never-ending curiosity about unresolved matters.

I now quote from my letter of acceptance of the 2006 Marshall Urist Awarded Lecture, bestowed upon me by Professors Reddi and Vukicevic, honorary president and president, respectively, of the Sixth International Conference on Bone Morphogenetic Proteins, Dubrovnik, Croatia:

I then drove to the Medical School of the University and climbed into the archives of the medical library to pick up the 1965 *Science* volume. When reading the summary, I felt enthralled and sat in the dusky and dusty room and avidly read the entire paper and just understood my re-location to Africa and the access to non-human primates *Papio ursinus* and with the animal model the major discoveries on the osteogenic molecular signals of the TGF-β superfamily not confined to the bone morphogenetic proteins (BMPs), but extending to several other pleiotropic members of the family, in particular, the three mammalian transforming growth factor-β (TGF-β) isoforms.

It went without saying that my doctoral studies needed to be perforce centered on the induction of bone formation in non-human primates (*Papio ursinus*). When working on my doctoral thesis, I was often confronted—as a clinician still struggling to understand the complexities of the induction processes—with some difficulties of interpretation of different articles on the bone induction cascade, and I mentioned to Tracey that I would have liked to discuss some matters with Dr. Urist himself to clear any possible misunderstandings. Tracey smiled and said, "Sure, why don't you phone him?" One late afternoon in 1986, while we were living in an apartment close to Hillbrow in Johannesburg, Tracey asked if I had phoned Dr. Urist. I guessed he wouldn't be easy to get a hold, also being a professor of orthopedic surgery. But Tracey went to the phone and, still smiling, asked me for the phone number of the Bone Research Laboratory at the University of California, Los Angeles (UCLA). She then phoned, and after some talking in fluent English (how I envied her capacity to speak English so well and fluently), she suddenly proffered the phone, saying, "Come here, Ugo. Dr. Urist is on the line. Talk to him." And so I did, and eventually 20 years later, I was given the opportunity and unique privilege to address the Marshall Urist Awarded Lecture during the opening ceremony of the Sixth Bone Morphogenetic Proteins Conference in Dubrovnik, Croatia. When establishing the Bone Research Laboratory at the medical school of the university in 1994, it went without saying that I wanted to keep the name of my laboratory in honor of Marshall Urist and in honor of his discoveries at the then Bone Research Laboratory at UCLA in the United States.

A grand change in my historical perspective on "*Bone: Formation by Autoinduction*" (Urist 1965), and indeed a grand challenge that has since affected my scientific thinking and research output, came along in June 1993 when I was honored as invited speaker at the International Symposium of Comprehensive Management of Craniofacial Anomalies—State of the Art, University of Aarhus, Denmark, where I presented a paper titled "Initiation of Craniofacial Bone Regeneration in Primates by Bone Morphogenetic Proteins."

I was privileged enough in the 1980s to have had the chance to learn the complete overview and perspectives of the phenomenon of the induction of bone formation when working in the laboratories of Dr. Reddi, together with his staff at the National Institutes of Health (NIH). All of us were thinking BMPs, BMPs, and nothing but BMPs, and like many scientists around the globe, we were focused on finally purifying the so-called osteogenic activity that was known to be present in the crude demineralized bone matrix after the classic experiments published in *Science* by Urist. The crude demineralized bone matrix obviously contains morphogenetic signals, or morphogens, first defined by Turing as "forms generating substances" (Turing 1952), that are endowed with the striking prerogative of setting into motion the induction of bone formation (Urist 1965; Reddi and Huggins 1972).

Hard work and continuous extraction and purification of large amounts of either bovine or baboon bone matrices rewarded the Bone Cell Biology Section in the United States and the Bone Research Laboratory in South Africa with important papers on the isolation to homogeneity of osteogenin, a bone morphogenetic protein, and its biological activity in rodents (Luyten et al. 1989) and non-human primates (*P. ursinus*) (Ripamonti et al. 1992, 1994). I grew up then as an osteogenin and later as a BMP man since all of us in both laboratories were working hard on how to apply BMPs or selected recombinant human BMPs (hBMPs) for either *in vitro* studies or *in vivo* studies in rodents and in the Chacma baboon (*P. ursinus*) (Reddi 1994, 2000; Ripamonti and Vukicevic 1995; Ripamonti et al. 1996b, 2000b), finally translating highly purified naturally derived osteogenic fractions in clinical contexts (Ripamonti and Ferretti 2002; Ferretti and Ripamonti 2002). Using purified protein preparations from baboon bone matrices and autolysed antigen-extracted allogeneic bone matrix preparations, the Bone Research Laboratory in South Africa provided the first unequivocal morphological evidence of bone formation by induction in heterotopic extraskeletal and orthotopic calvarial sites of adult Chacma baboons (Ripamonti 1991; Ripamonti et al. 1992), including periodontal tissue regeneration (Ripamonti et al. 1994).

In a tight collaborative research effort between the Bone Cell Biology Section at the National Institutes of Health and the University of the Witwatersrand, Johannesburg, with H.A. Reddi of NIH (Bethesda, MD) and Laura Yates, Noreen Cunningham, and Shu-Shan Ma, we purified osteogenin, then thought to be BMP-3 (Luyten et al. 1989), to homogeneity from baboon bone matrices. Final purification to homogeneity was obtained by electroendosmotic elution by means of sodium dodecyl sulfate (SDS) polyacrylamide gel electrophoresis (PAGE) (Ripamonti et al. 1992). This resulted in a single band on SDS-PAGE with a corresponding apparent molecular weight of 32 kDa with biological activity in rats (Ripamonti et al. 1992). The bone regeneration potential was investigated in non-healing calvarial defects surgically prepared in adult baboons (Ripamonti et al. 1992). At the Bone Research Laboratory in South Africa, highly purified bovine osteogenic fractions, after additional affinity chromatography on heparin–Sepharose, were used to treat large mandibular defects in human patients (Ripamonti and Ferretti 2002; Ferretti and Ripamonti 2002), translating the bone induction principle (Urist et al. 1967) into clinical contexts. Purification and implantation were televised worldwide by *Beyond 2000* in 1997.

An account of the incredible amount of research work we all were doing while at the NIH Bone Cell Biology Section in the late 1980s and beginning of the 1990s, by several scientists from all over the world, centered on the magnetic vibrant personality of Hari Reddi, was published a few years ago in *Science in Africa* as a feature article on the geometric induction of bone formation (Ripamonti 2012). Significantly, Laura, attached to the then embryonic Bone Research Laboratory in South Africa, excelled in purifying batches of osteogenin to homogeneity, further contributing to my quest of the induction of bone formation impeccably eluting highly active, purified, naturally derived osteogenic fractions that were loaded onto macroporous constructs of coral-derived bioreactors, constructing a chromatographic bed for adsorption of extracted and uploaded BMPs *in vivo* (Ripamonti et al. 1993).

I often pause during current experimentation or when composing manuscripts to think about the many hours of hard work into the nights during that scientifically highly rewarding period at NIH, frequently flying back and forth from the Dental Research Institute in South Africa. We were all focused on the potential translation into clinical contexts of the purified and then cloned BMPs, dreaming that our work at the laboratory bench would soon be translated into clinical contexts to treat crippling disorders of the human skeleton (Urist 1968).

Excited by the perspective to contribute to the common understanding of Nature, and somehow enthralled and mesmerized by our own discoveries and continuous progress, Laura and I were continuously talking science and planned *in vitro* studies for growing osteoblastic-like cells, later to be combined and grown into coral-derived macroporous constructs to engineer a bone bioreactor for potential application and transplantation in vivo (Ripamonti 2012; Ripamonti et al. 2012). Laura's experiments were prototypes in engineering a biological construct with embedded geometric cues for the induction of bone formation in extraskeletal heterotopic sites (Ripamonti 2012; Ripamonti et al. 2012). We did indeed later construct a heterotopic bone graft in the chest muscular tissue using coral-derived macroporous constructs that were later transplanted in a human mandibular defect (see Figure 2.7) (Heliotis et al. 2006).

A dichotomy, however, did happen in the 1990s when, by serendipity, my scientific understanding and research output turned toward the need to reevaluate the induction of bone formation after discovering a redundancy of other morphogens, homologous yet molecularly different from the bone morphogenetic proteins. The turning point was during the proceedings of the above-mentioned conference in Aarhus, where I had the privilege to meet with Prof. Birte Melsen and Dr. Carlos Bosh, a PhD student visiting from Spain. Carlos was preparing his doctoral studies on the biological activity of the first mammalian transforming growth factor-$\beta_1$ (TGF-$\beta_1$) isoform after implantation in bony defects. Our meeting suggested to him the use of the Chacma baboon as an experimental model for his doctoral studies. I was not in any way keen to test the biological activity of any TGF-$\beta$ isoform in our baboon model at the Central Animal Services of the university. It dawned on me, however, that it would be appropriate to host a visiting professor from overseas just after the official opening of the Bone Research Laboratory at the medical school of the university, and to show the world that implantation of doses of recombinant hTGF-$\beta_1$ would not induce bone formation when implanted in the *rectus abdominis* muscle of *P. ursinus*, where there is no bone, the acid test to prove that a soluble molecular signal and/or a device is indeed endowed with the striking prerogative of *de novo* generating new bone formation (Ripamonti 2012).

In further discussion with Carlos, I told him that I would be interested in hosting him at the University of the Witwatersrand, Johannesburg, but that doses of hTGF-$\beta_1$ would be far too expensive for my university's allocated budget. Carlo smiled and said that the University of Aarhus had a very good relationship with the University of California and Genentech, Inc., the colossus biotechnology company owning patents on the hTGF-$\beta_1$ isoform, and thus producing the recombinant protein. In January 1994, Dr. Carlos Bosh flew from Denmark, bringing with him milligram amounts of recombinant hTGF-$\beta_1$.

The recombinant morphogen was promptly implanted in calvarial defects 25 mm in diameter and in heterotopic intramuscular sites of four Chacma baboon (*P. ursinus*). At euthanasia, we were confronted with seemingly opposite results between the orthotopic and heterotopic sites. Calvarial specimens showed limited, if any, induction of bone formation, particularly on day 30, with, however, interesting islands of chondrogenesis within the newly formed bone close to the craniotomies' margins (Ripamonti et al. 1996a). Heterotopically, however, there was the induction of massive bone across the implanted *rectus abdominis* muscle. The implantation scheme and the distance between implants of insoluble collagenous bone matrix reconstituted with recombinant human osteogenic protein-1 (hOP-1) or hTGF-$\beta_1$ are described in Chapter 2 and particularly Chapter 6 and in Ripamonti et al. (1997). The available data indicated beyond doubt that the hTGF-$\beta_1$ was endowed with the striking capacity of inducing endochondral bone in the *rectus abdominis* muscle of the baboon. After several discussions among us at the Bone Research Laboratory, we eventually concluded that first, recombinant hTGF-$\beta_1$ is *per se* an initiator of the bone induction cascade, and second, hTGF-$\beta_1$ synergizes with hOP-1 to induce rapid and substantial bone formation in heterotopic sites of *P. ursinus*, providing evidence for a novel function of hTGF-$\beta_1$ in primates and the scientific basis for synergistic molecular therapeutics for the rapid generation of cartilage and bone (Ripamonti et al. 1997). However, I was still perturbed by the fact that against all scientific and commercial dogmas, the hTGF-$\beta_1$ suddenly turned out to be inductive, but in a non-human primate model. I deemed it necessary to reimplant doses of the hTGF-$\beta_1$ protein singly or in combination with hOP-1 in other animals to resolve the definition of a new function of the molecular isoform in primates. The implantation schemes are described in Chapters 2 and 6 and reported in detail in the original paper that first described the novel functions of the isoform as an initiator of bone formation in heterotopic non-human primate sites and triggering, together with hOP-1, a long range of biological induction across the rectus abdominis muscle, resulting in the synergistic induction of bone formation at a specific ratio of 20:1 by weight of the two recombinant morphogens (Ripamonti et al. 1997).

The submission of the manuscript to the *Journal of Bone and Mineral Research* was then completed, followed by an editorial letter that would have set back the experimentation, requesting several additional animals to prove the point that the hTGF-$\beta_1$ protein was indeed inductive. I was still meditating the editorial letter with the associated review comments, trying hard to prepare a rebuttal, when, unexpectedly, I received a letter from the office of the editor of the *Journal of Bone and Mineral Research*, which I am happy to transcribe verbatim below:

Dear Dr Ripamonte

Re: JBMR #D702 047

Your paper referenced above has been judged interesting and potentially acceptable for publication in the *Journal of Bone and Mineral Research*. The recommendations of Reviewer #1 and #2 were in conflict. The editors consulted with an acknowledged expert in the field because of the extensive nature of the experiments recommended by Reviewer #1. It was the opinion of the adjudicating scientist, with which we agree, that performing further experiments on the scale recommended by reviewer #1 *is not necessary if the following points are addressed in a revision.*

The minor points were, of course, rapidly addressed, and the paper was published later in the same year (Ripamonti et al. 1997). The rather unique work on the induction of bone formation by the hTGF-$\beta_1$ isoforms was followed by an investigation into the biological activity of platelet-derived porcine transforming growth factor-$\beta_1$ (pTGF-$\beta_1$) (Duneas et al. 1998). This body of experimentation confirmed both the inductive activity of the first mammalian TGF-$\beta$ isoform and the synergistic induction of bone formation with identical doses of the hOP-1 (Ripamonti et al. 1997; Duneas et al. 1998).

More work followed, testing the biological activity of the hTGF-$\beta_2$ isoform, which was later published in *Growth Factors* (Ripamonti et al. 2000a). The Bone Research Laboratory found itself the only laboratory in the world to be actively working on the bone induction prerogative of the mammalian TGF-$\beta$ isoforms (Ripamonti 2003, 2004; Ripamonti et al. 2004; Ripamonti et al. 2006; Ripamonti et al. 2010). We felt, however, that the scientific or commercial worlds were not seriously appraising our scientific efforts and publications thereafter. After meeting the editor in chief of *Biomaterials* in 2003 in Cape Town, I had the opportunity to prepare my first leading opinion paper for *Biomaterials*, which was published in 2006 (Ripamonti 2006a). Later, much morphological and molecular work went into the preparation of our first manuscript on the induction of bone formation by the hTGF-$\beta_3$ isoform, which eventually was published by the *Journal of Cellular and Molecular Medicine* (Ripamonti et al. 2008). Yet in spite of our continuous efforts in testing the hTGF-$\beta_3$ osteogenic device in both non-human and human primates (Klar et al. 2014; Ripamonti et al. 2014, 2015b), the inductive potential of the three mammalian TGF-$\beta$ isoforms has only been studied and confirmed in our laboratories, and never confirmed by any laboratory outside ours.

My experience with the induction of bone formation by the mammalian TGF-$\beta$ isoforms has been the beginning of a fascinating scientific and intellectual journey across cellular induction, redundancy, speciation, molecular evolution, stem cell differentiation, dedifferentiation, and the induction of large masses of corticalized bone by the hTGF-$\beta_3$ isoform, as shown in the iconographic plates of the various chapters in this volume. The turning point happened when, as recipient of the Marshall Urist Awarded Lecture in 2006, Dubrovnik, Croatia, I unveiled to the scientific community the rapid induction of bone formation by the hTGF-$\beta_3$ osteogenic device (Ripamonti 2006b), also reporting on the *restitutio ad integrum* of mandibular defects in *P. ursinus* by day 30 with doses of the hTGF-$\beta_3$ protein (Ripamonti 2006b). Dr. Peter ten Dijke, together with Dr. Aris Economides, has been the only colleague to truly help to significantly unravel the mechanistic insights of the induction of bone formation by the hTGF-$\beta_3$ protein in non-human primates (Klar et al. 2014; Ripamonti et al. 2014, 2015b). I am sorry Peter was not able to find the time to write an invited chapter on the molecular re-evaluation of the three mammalian TGF-$\beta$ isoforms. As the author of this volume on the powerful inductive activity of the hTGF-$\beta_3$ isoform, I was saddened by the difficulties Hari A. Reddi has had in providing introductory statements for it. The writer is also missing the contribution of Laura, who would have contributed an important chapter to this volume.

Special thanks go to Ruqayya Parak for the preparation of undecalcified sections using the Exakt diamond saw grinding and polishing system. Ruqayya is certainly one of the best, if not the best, technologist in the southern hemisphere, and to her, special recognition is offered. Special thanks also go to Sindisiwe Shangase of the School of Oral Health Sciences for her vision to allow me precious time to compose and complete the task of writing a scientific book; special recognition is also offered to her.

An important word of thanks must go to Novartis AG, Zurich, Switzerland, which provided aliquots of the recombinant hTGF-$\beta_3$ protein.

A special word of thanks to CRC Press and its staff, who had to handle an often recalcitrant author. Special thanks to at least three persons at CRC Press, Taylor & Francis: Leong Li-Ming (梁丽明), acquisitions editor; Hayley Ruggieri, project coordinator, editorial project development; and Chuck R. Crumly, senior acquisitions editor.

I have no time to regret or feel saddened about what would been if additional chapters had been included. My laboratory is now involved in molecularly and morphologically dissecting a time course study that has mechanistically shown the induction of bone formation starting at day 17 after heterotopic implantation of doses of the hTGF-$\beta_3$ isoform, recruiting several stem cells at the site of implantation, later transformed into mineralized constructs at the periphery of the implanted coral-derived macroporous bioreactors (Ripamonti et al. 2015a).

We have shown that hTGF-$\beta_3$-treated bioreactors significantly upregulate *RUNX-2* and *Osteocalcin* on day 15, controlling the differentiation of progenitor stem cells into the osteoblastic phenotype. Our data show that the rapid induction is possible because the recombinant protein reprograms recruited differentiated myoblastic cells and pericytes into highly active secreting osteoblasts in the *rectus abdominis* striated muscle of *P. ursinus* (Klar et al. 2014; Ripamonti et al. 2014, 2015b).

In concluding, many scientists and research technologists are kindly acknowledged in the various chapters collected in this volume; they were instrumental for the birth of this contribution. Special thanks go to the Chacma baboon (*P. ursinus*), which has constantly and continuously provided a series of unique positive microenvironments for the induction of bone formation to occur. Special thanks to Barbara van den Heever, who initiated undecalcified histology using tungsten carbide blades and provided the writer with sections at 3 μm and less, now published worldwide. The molecular discoveries of the induction of bone formation by the hTGF-$\beta_3$ could not have been possible without the hard work of Roland Klar, then PhD student in my laboratories and now post-doctoral fellow in both the Bone Research Laboratory and the Laboratory of Molecular and Cellular Biology, Department of Internal Medicine, headed by Dr Raquel Duarte. The continuous hard work of Roland Klar, Therese Dix-Peek, Caroline Dickens and Raquel Duarte, the molecular biology team, has significantly contributed to unravel the mysteries of the induction of bone formation in primates.

Finally, I wish to dedicate this work to my daughter, Daniella Bella, who has inspired my life, and thus my scientific productivity, when trying hard to rebuild myself after shattering life events, and I wish the readers of this volume to grasp the message that *never* is cast in iron, even in science.

## References

DUNEAS, N., CROOKS, J., RIPAMONTI, U. (1998). Transforming growth factor-$\beta_1$: Induction of bone morphogenetic protein genes expression during endochondral bone formation in the baboon, and synergistic interaction with osteogenic protein-1 (BMP-7). *Growth Factors* 15, 259–77.

FERRETTI, C., RIPAMONTI, U. (2002). Human segmental mandibular defects treated with naturally-derived bone morphogenetic proteins. *J Craniofac Surg* 13, 434–44.

HELIOTIS, M., LAVERY, K.M., RIPAMONTI, U., TSIRIDIS, E., DI SILVIO, L. (2006). Transformation of a prefabricated hydroxyapatite/osteogenic protein-1 implant into a vascularised pedicled bone flap in the human chest. *Intern J Oral Maxillofac Surg* 35, 265–69.

KLAR, R.M., DUARTE, R., DIX-PEEK, T., RIPAMONTI, U. (2014). The induction of bone formation by the recombinant human transforming growth factor-$\beta_3$. *Biomaterials* 35(9), 2773–88. http://dx.doi.org/10.1016/j.biomaterials.2013.12.062.

LUYTEN, F.P., CUNNINGHAM, N.S., MA, S., MUTHUKUMARAN, N., HAMMONDS, R.G., NEVINS, W.B., WOODS, W.I., AND REDDI, A.H. (1989). Purification and partial amino acid sequence of osteogenin, a protein initiating bone differentiation. *J Biol Chem* 264(23), 13377–80.

REDDI, A.H. (1994). Symbiosis of biotechnology and biomaterials: Applications in tissue engineering of bone and cartilage. *J Cell Biochem* 56(2), 192–95.

REDDI, A.H. (1997). Bone morphogenesis and modeling: Soluble signals sculpt osteosomes in the solid state. *Cell* 89, 159–61.

REDDI, A.H. (2000). Morphogenesis and tissue engineering of bone and cartilage: Inductive signals, stem cells, and biomimetic biomaterials. *Tissue Eng* 6(4), 351–59.

REDDI, A.H., and HUGGINS, C. (1972). Biochemical sequences in the transformation of normal fibroblasts in adolescent rats. *Proc Natl Acad Sci USA* 69(6), 1601–5.

RIPAMONTI, U. (1991). Bone induction in non-human primates: An experimental study on the baboon. *Clin Orthop Relat Res* 269, 284–94.

RIPAMONTI, U. (2003). Osteogenic proteins of the transforming growth factor-$\beta$ superfamily. In H.L. Henry and A.W. Norman (eds.), *Encyclopedia of Hormones.* Academic Press, San Diego, CA, pp. 80–86.

RIPAMONTI, U. (2004). Soluble, insoluble and geometric signals sculpt the architecture of mineralized tissues. *J Cell Mol Med* 8(2), 169–80.

RIPAMONTI, U. (2006a). Soluble osteogenic molecular signals and the induction of bone formation. *Biomaterials* 27, 807–22.

RIPAMONTI, U. (2006b). The Marshall Urist Awarded Lecture. Bone: Formation by autoinduction. In S. Vukicevic and A.H. Reddi (eds.), *Proceedings of the 6th International Conference on BMPs*, Dubrovnik, Croatia.

RIPAMONTI, U. (2012). The concavity: The "shape of life" and the control of bone differentiation—feature paper—Science in Africa; Ripamonti U., Roden L., Renton L., Klar R., Petit J.-C. The influence of geometry on bone: formation by autoinduction. http://www.scienceinafrica.co.za/2012/Ripamonti_bone.htm.

RIPAMONTI, U., BOSCH, C., VAN DEN HEEVER, B., DUNEAS, N., MELSEN, B., EBNER, R. (1996a). Limited chondro-osteogenesis by recombinant human transforming growth factor-$\beta_1$ in calvarial defects of adult baboons (*Papio ursinus*). *J Bone Miner Res* 11, 938–45.

RIPAMONTI, U., CROOKS, J., MATSABA, T., TASKER, J. (2000a). Induction of endochondral bone formation by recombinant human transforming growth factor-$\beta_2$ in the baboon (*Papio ursinus*). *Growth Factors* 17(4), 269–85.

RIPAMONTI, U., DICKENS, C., DIX-PEEK, T., PARAK, R., KLAR, M.R. (2015a). In preparation.

RIPAMONTI, U., DIX-PEEK, T., PARAK, R., MILNER, B., DUARTE, R. (2015b). Profiling bone morphogenetic proteins and transforming growth factor-$\beta$s by hTGF-$\beta_3$ pre-treated coral-derived macroporous constructs: The power of one. *Biomaterials* 49, 90–102.

RIPAMONTI, U., DUARTE, R., FERRETTI, C. (2014). Re-evaluating the induction of bone formation in primates. *Biomaterials* 35, 9407–22.

RIPAMONTI, U., DUNEAS, N., VAN DEN HEEVER, B., BOSCH, C., CROOKS, J. (1997). Recombinant transforming growth factor-$\beta_1$ induces endochondral bone in the baboon and synergizes with recombinant osteogenic protein-1 (bone morphogenetic protein-7) to initiate rapid bone formation. *J Bone Miner Res* 12, 1584–95.

RIPAMONTI, U., FERRETTI, C. (2002). Mandibular reconstruction using naturally-derived bone morphogenetic proteins: A clinical trial report. In T.S. Lindholm (ed.), *Advances in Skeletal Reconstruction Using Bone Morphogenetic Proteins*. Singapore: World Scientific, pp. 277–89.

RIPAMONTI, U., FERRETTI, C., HELIOTIS, M. (2006). Soluble and insoluble signals and the induction of bone formation: Molecular therapeutics recapitulating development. *J Anat* 209, 447–68.

RIPAMONTI, U., HELIOTIS, M., VAN DEN HEEVER, B., REDDI, A.H. (1994). Bone morphogenetic proteins induce periodontal regeneration in the baboon (*Papio ursinus*). *J Periodont Res* 29, 439–45.

RIPAMONTI, U., KLAR, R.M., RENTON, L.F., FERRETTI, C. (2010). Synergistic induction of bone formation by hOP-1 and TGF-β3 in macroporous coral-derived hydroxyapatite constructs. *Biomaterials* 31(25), 6400–10.

RIPAMONTI, U., MA, S., CUNNINGHAM, N., YATES, L., REDDI, A.H. (1992). Initiation of bone regeneration in adult baboons by osteogenin, a bone morphogenetic protein. *Matrix* 12, 202–12.

RIPAMONTI, U., RAMOSHEBI, L.N., PATTON, J., MATSABA, T., TEARE, J., RENTON, L. (2004). Soluble signals and insoluble substrata: Novel molecular cues instructing the induction of bone. In E.J. Massaro and J.M. Rogers (eds.), *The Skeleton.* Humana Press, Totowa, New Jersey, pp. 217–27.

RIPAMONTI, U., RAMOSHEBI, L.N., TEARE, J., RENTON, L., FERRETTI, C. (2008). The induction of endochondral bone formation by transforming growth factor-β3: Experimental studies in the non-human primate *Papio ursinus. J Cell Mol Med* 12(3), 1029–48.

RIPAMONTI, U., RODEN, L.C., RENTON, L.F. (2012). Osteo-inductive hydroxyapatite-coated titanium implants. *Biomaterials* 33, 3813–23.

RIPAMONTI, U., VAN DEN HEEVER, B., CROOKS, J., TUCKER, M.M., SAMPATH, T.K., RUGER, D.C. (2000b). Long term evaluation of bone formation by osteogenic protein-1 in the baboon and relative efficacy of bone-derived bone morphogenetic proteins delivered by irradiated xenogeneic collagenous matrices. *J Bone Miner Res* 15, 1798–809.

RIPAMONTI, U., VAN DEN HEEVER, B., SAMPATH, T.K., TUCKER, M.M., RUEGER, D.C., REDDI, A.H. (1996b). Complete regeneration of bone in the baboon by recombinant human osteogenic protein-1 (bone morphogenetic protein-7). *Growth Factors* 13, 273–89.

RIPAMONTI, U., VUKICEVIC, S. (1995). Bone morphogenetic proteins: From developmental biology to molecular therapeutics. *South Afr J Sci* 91, 277–80.

RIPAMONTI, U., YEATES, L., VAN DEN HEEVER, B. (1993). Initiation of heterotopic osteogenesis in primates after chromatographic adsorption of osteogenin, a bone morphogenetic protein, onto porous hydroxyapatite. *Biochem Biophys Res Commun* 193, 509–17.

TURING, A.M. (1952). The chemical basis of morphogenesis. *Philos R Soc Lond* 237, 37.

URIST, M.R. (1965). Bone: Formation by autoinduction. *Science* 150(698), 893–99.

URIST, M.R. (1968). The reality of a nebulous enigmatic myth. *Clin Orthop Relat Res* 59, 3–5.

URIST, M.R., SILVERMAN, B.F., BURING, K., DUBUC, F.L., AND ROSENBERG, J.M. (1967). The bone induction principle. *Clin Orthop Relat Res* 53, 243–83.

# CHAPTER TWO

# Regenerative Medicine, the Induction of Bone Formation, Bone Tissue Engineering, and the Osteogenic Proteins of the Transforming Growth Factor-β Supergene Family

*Ugo Ripamonti*

Bone Research Laboratory, School of Oral Health Sciences, Faculty of Health Sciences, University of the Witwatersrand, Johannesburg, Parktown, South Africa

## 2.1 Regenerative Medicine and Bone: Formation by Autoinduction

Regenerative medicine is the grand multidisciplinary challenge of molecular, cellular, and evolutionary biology, requiring the integration of molecular and tissue biology, tissue engineering, developmental biology, and experimental surgery to explore how to trigger *de novo* induction of tissues and organs of the mammalian body, with the ultimate goal of inducing *de novo* and *ex novo* tissues and organs in man (Sampath and Reddi 1981, 1983; Khouri et al. 1991; Reddi 1994, 2000; Viola et al. 2003).

A central question in developmental biology, and thus tissue engineering and regenerative medicine, is the molecular basis of pattern formation, tissue induction, and morphogenesis (Reddi 1984; Lander 2007; De Robertis 2008). The pursuit of understanding pattern formation, the attainment of tissue form, and function or morphogenesis (Reddi 1984) has been vigorously pursued for decades and has been the source of significant advances and even greater frustration (Ripamonti et al. 2009). It has dramatically carved the definition (Turing 1952)—and later the discovery—of morphogenetic signals or morphogens, which are now at the very crux of tissue induction and morphogenesis (Lander 2007; Kerszberg and Wolpert 2007; Urist 1965; Urist et al. 1967). Morphogenesis, the genesis of form, is induced by morphogens (Turing 1952). As a prerequisite to tissue differentiation and morphogenesis, there must be several signaling molecules, or morphogens, first defined by Turing (1952) as "forms generating substances," that initiate tissue induction and morphogenesis. Morphogens interact with specific cell surface receptors on responding cells, which may or may not be stem cells, to initiate the ripple-like cascade of pattern formation, tissue induction, and the attainment of tissue form and function, or morphogenesis (Reddi 1997, 2000; Ripamonti et al. 2004).

The aim of this introductory chapter to the induction of bone formation by the three mammalian transforming growth factor-$\beta$ (TGF-$\beta$) isoforms, and particularly the rapid induction of bone formation by the third mammalian TGF-$\beta$ isoform, is to convey a concise perspective on the induction of bone formation and its applications in the context of regenerative medicine and bone tissue engineering. The rationale of linking bone formation by induction to regenerative medicine is based on the discovery of the vast pleiotropic activities of the osteogenic proteins of the TGF-$\beta$ supergene family (Reddi 2000; Ripamonti 2003), which provide soluble osteogenic molecular signals endowed with the striking prerogative of initiating *de novo* bone formation in heterotopic extraskeletal sites of a variety of animal models, including non-human primates (Ripamonti 2005, 2006; Ripamonti et al. 2004, 2005). At the same time, the now recently labeled body morphogenetic proteins (BMPs) (Reddi 2005) indicate that they are pleiotropic regulatory morphogens controlling a vast array of biological functions outside the mere osseous domain (Reddi 2000, 2005; Ripamonti et al. 2005; Ripamonti 2006).

We have thus learned that the induction of bone formation requires three key components (Reddi 2000; Ripamonti et al. 2000b, 2004): soluble osteogenic molecular signals, responding stem cells, and insoluble signals or substrata, upon which differentiating mesenchymal stem cells erect *"Bone: Formation by autoinduction"* (Urist 1965; Reddi and Huggins 1972; Reddi 2000). The induction of bone formation by recombining or reconstituting osteogenic soluble molecular signals with insoluble signals or substrata has been pivotal for setting the rules of the tissue engineering paradigm (Reddi 2000). Ultimately, the tissue engineering paradigm is the induction of tissue morphogenesis by combinatorial molecular protocols whereby soluble molecular signals are recombined to insoluble signals or substrata to effectively reconstitute the biological activity of the soluble osteogenic molecular signals (Figure 2.1) (Sampath and Reddi 1981, 1983; Ripamonti and Reddi 1995; Khouri et al. 1991; Reddi 1994, 2000; Ripamonti et al. 2001, 2004).

The induction of bone formation has evolved as a prototype of the tissue engineering paradigm, continuously developing the concept and further expanding the depth and breadth of the novel developmental, biological, and surgical concepts (Viola et al. 2003; Reddi 2000, 2005). Importantly, classic studies on the induction of bone formation have set the critical rules that govern tissue engineering and regenerative medicine (Sampath and Reddi 1981, 1983; Ripamonti and Reddi 1995; Khouri et al. 1991; Reddi 2000; Ripamonti et al. 2001, 2004).

As stated by Reddi in his incisive minireview in *Cell*, "bone, like all matter, is in both a soluble and a solid state and there is a *continuum* between the soluble and solid states that is regulated by signals in solution interacting with the insoluble extracellular matrix" (Reddi 1997). The biochemical problem of the extracellular matrix in the solid state (Reddi 1997) has been resolved by the chaotropic dissociative extraction of the bone matrix, yielding in an insoluble and a soluble component from which putative osteogenic proteins tightly bound to the organic highly mineralized extracellular matrix of bone could be extracted and solubilized (Sampath and Reddi 1981, 1983; Ripamonti and Reddi 1995). This has resulted in the isolation, purification, and characterization of an entirely new family of protein initiators (Wozney et al. 1988; Celeste et. al 1990; Sampath et al. 1990; Reddi 2000; Ripamonti 2006) masterminding axial patterning, symmetric and asymmetric tissue development, morphogenesis, and organogenesis of several tissues and organs from the fruit fly *Drosophila melanogaster* to *Homo sapiens* and defined as the body morphogenetic proteins or bone morphogenetic proteins (BMPs) (Figure 2.2) (Reddi 2005); BMPs are members of the TGF-β supergene family (Wozney et al. 1988; Celeste et. al 1990; Reddi 2000; Ripamonti 2006).

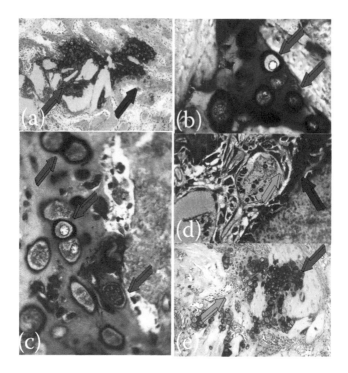

**FIGURE 2.1** Tissue induction and morphogenesis by combinatorial molecular protocols whereby 0.1–0.5 μg of naturally derived, highly purified baboon osteogenic protein fractions purified greater than 50,000-fold was reconstituted with 25 mg of rat allogeneic insoluble collagenous bone matrix as carrier and implanted in the subcutaneous space of Long–Evans rats (Ripamonti et al. 1992). Protein extracts were sequentially purified on hydroxyapatite Ultro-Gel adsorption and heparin–Sepharose affinity chromatography, followed by gel filtration chromatography on tandem S-200 Sephacryl columns (Ripamonti et al. 1992). Tissue specimens harvested on day 12 after heterotopic implantation were processed for undecalcified histology and embedded in historesin (Ripamonti et al. 1992). (a) Low-power microphotograph showing the induction of chondrogenesis (magenta arrow) with the induction of bone formation (dark blue arrow) in close proximity to the collagenous matrix as carrier. (b) Chondrogenesis attached to the insoluble carrier matrix with hypertrophic chondrocytes (magenta arrows) within the newly secreted cartilaginous matrix. (c) Fragmentation of the cartilaginous matrix and hypertrophic chondrocytes (magenta arrows) during tissue induction and morphogenesis by highly purified, naturally derived osteogenic protein fractions. (d) Induction of angiogenesis in close contact with the osteoblastic compartment along the implanted insoluble carrier (dark blue arrow). Note the hypertrophic hyperchromatic endothelial cells seemingly detaching from the vascular compartment (light blue arrow), migrating toward the osteoblastic compartment as a continuous source of progenitor cells for the induction of osteogenesis in angiogenesis (Ripamonti 2006; Ripamonti et al. 1992, 2006, 2007). (e) Induction of chondrogenesis (magenta arrow) and osteogenesis (light blue arrow) by 2 μg of recombinant human osteogenic protein-1 (hOP-1) when reconstituted with collagenous matrix and implanted in the subcutaneous space of the rat. Undecalcified sections in historesin cut at 3 μm and stained with Toluidine blue.

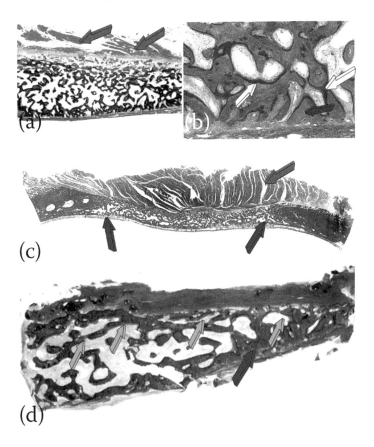

**FIGURE 2.2**    *"Bone: Formation by autoinduction"* (Urist 1965) in the non-human primate *Papio ursinus* and in the primate *Homo sapiens* using highly purified baboon (a–c) and bovine (d) osteogenic protein fractions purified greater than 50,000-fold by sequential chromatography on hydroxyapatite Ultro-Gel adsorption and heparin–Sepharose affinity chromatography, followed by gel filtration chromatography on tandem S-200 Sephacryl columns (Ripamonti et al. 1992). (a) Induction of bone formation by 280 μg of naturally derived, highly purified osteogenic fractions combined with allogeneic insoluble collagenous bone matrix on day 30 after implantation in nonhealing calvarial defects of *P. ursinus* covered by the pericranium and the temporalis muscle (magenta arrows) (Ripamonti et al. 1992, 2005; Ripamonti 2006). (b) High-power view highlighting newly formed mineralized bone (dark blue arrow) covered by osteoid seams (white arrows) populated by contiguous osteoblasts. (c) Induction of bone formation across the calvarial defect 90 days after implantation of 280 μg of naturally derived, highly purified baboon osteogenic fractions combined with allogeneic insoluble collagenous bone matrix with regeneration of the calvarial defect) (dark blue arrows). (d) Highly purified bovine osteogenic protein fractions were reconstituted with human demineralized bone matrix and implanted in mandibular defects of human patients (Ripamonti and Ferretti 2003). (d) Biopsy section showing mineralized newly formed bone (dark blue arrow) covered by osteoid seams (light blue arrows) populated by osteoblastic cells. Undecalcified blocks embedded in K-Plast with sections cut at 6 μm stained free-floating with Goldner's trichrome.

The challenging problem of the induction of bone formation in non-human and human primates (Urist 1968) has stimu-
lated the Bone Research Laboratory of the University of the
Witwatersrand, Johannesburg, to develop experimental animal
models using adult non-human primates of the species *Papio
ursinus* that share similar bone physiology and remodeling with
man (Schnitzler et al. 1993). Importantly, *P. ursinus* is a suitable model for the study of postmenopausal bone loss consis-
tent with and resembling human postmenopausal osteoporosis
(Dal Mas et al. 2007), further highlighting similar osteonic
bone remodeling with man (Schnitzler et al. 1993).

The experimental work of Urist (1968; Urist and McLean 1952;
Urist et al. 1967, 1968), adding to the previously monumental
seminal work of Levander (1938, 1945; Levander and Willestaedt
1946) and Moss (1958), borrowing the term *induction* from
Spemann (1938) and Levander (1945), reported the reproducible evidence that the implantation of allogeneic demineralized
bone matrix (DBM) in heterotopic intramuscular sites of rodents
and lagomorphs and in a variety of orthotopic bone defects of
humans, resulted, as per his classic studies published in *Science*
1965, in *"Bone: Formation by autoinduction"* (Urist 1965).

The osteogenic soluble molecular signals of the TGF-β
supergene family, the BMPs, and uniquely, in primates only,
the three mammalian TGF-β isoforms (Ripamonti et al. 1997,
2000a, 2008, 2004; Ripamonti and Roden 2010; Klar et al.
2014) induce endochondral bone formation as a recapitulation of embryonic development (Figures 2.1 and 2.2) (Reddi
and Huggins 1972; Reddi 2000; Ripamonti et al. 2001, 2004;
Ripamonti 2006, 2007). Importantly, for tissue engineering in
clinical contexts, neither the solubilized proteins nor the insoluble signals or residues are active (Sampath and Reddi 1981,
1983); the reconstitution of the soluble and insoluble components of the bone matrix does, however, restore the osteogenic
activity of the solubilized and now combined soluble signals,
thus resulting in the induction of bone formation in the rodent
extraskeletal heterotopic bioassay (Figure 2.1) (Sampath and
Reddi 1981, 1983; Ripamonti and Reddi 1995; Reddi 2000;
Ripamonti et al. 2001, 2004).

The operational reconstitution of the soluble molecular signal with an insoluble signal or substratum was a key experiment
that provided a bioassay for *bona fide* initiators of endochondral
bone differentiation (Sampath and Reddi 1981, 1983; Ripamonti
and Reddi 1995) in heterotopic extraskeletal sites of animal
models (Figure 2.1), including non-human and human primates

in orthotopic skeletal sites (Figures 2.2 through 2.4) (Ripamonti et al. 1992, 1996, 2000b, 2001, 2004; Ripamonti 2004, 2009).

## 2.2 Induction of Bone Formation: Osteogenesis in Angiogenesis

In his classic studies on the role of vessels in angiogenesis, Trueta (1963) quoted the insights of von Haller, who made the "then extravagant suggestion that the vascular system was responsible for osteogenesis" (von Haller 1763). Trueta further quoted the insights of Keith, who has suggested that bone-forming cells are derived from the endothelium of the invading capillaries (Keith 1927). Trueta, while presenting an outstanding lucid and clear morphological vision of osteogenesis in angiogenesis, highlighted the progeny of the bone-forming cells or osteoblasts and of the bone resorptive cells or osteoclasts (Trueta 1963). This provided the first insights into the supramolecular assembly of the extracellular matrix of bone. Trueta hypothesized that there is a *syncytium* of the bone-forming cells and osteocytes connected to the capillary network *via* the canaliculae of the osteonic bone matrix, with embedded osteocytes cemented within the mineralized extracellular matrix of bone (Trueta 1963; Vukicevic et al. 1990). The syncytium of the bone matrix as a whole has indicated that the skeleton is an organ connected by the canaliculae of the bone matrix with embedded osteocytes (Trueta 1963).

Trueta, reviewing the theory of the induction of osteogenesis, remarked that a "local substance operates directly on the vascular system causing an angioblastic specific stimulation of the bone vessels," an unknown substance he named the vascular stimulating factor (VSF) (Trueta 1963); Trueta's VSF was the first description of the existence of the vascular endothelial growth factor (VEGF), a critical soluble signal secreted during the cascade of bone formation by induction and maintenance of the newly induced bone (Leung et al. 1989; Byrne et al. 2005; Carlevaro et al. 2000; Gerber et al. 1999; Folkman and D'Amore 1996).

Yet, long before the studies of von Haller (1763) and the lucid work of Trueta (1963), Aristotle (384–322 BC), as reported by Lanza and Vegetti (1971) and Crivellato et al. (2007), credited the forming blood vessels with a patterning scenario of organogenetic function during organogenesis. Aristotle further commented that the architectural patterning of vessel growth

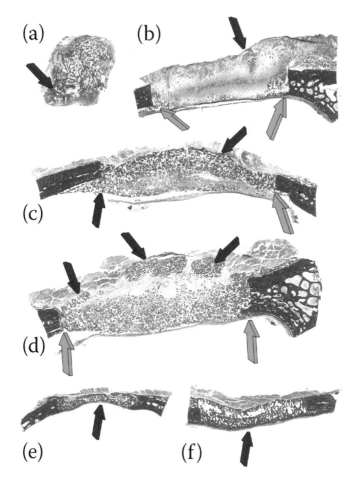

**FIGURE 2.3** Morphology of calvarial regeneration after implantation of doses of gamma-irradiated osteogenic devices of recombinant human osteogenic protein-1 (hOP-1) combined with bovine insoluble collagenous bone matrix as carrier (Ripamonti 2005). (a) Biological activity of hOP-1 in heterotopic intramuscular sites of *Papio ursinus*: induction of a large mineralized corticalized ossicle (dark blue arrow) in the *rectus abdominis* muscle of the adult non-human primate *P. ursinus* upon intramuscular implantation of 500 µg of hOP-1 per gram of insoluble bovine collagenous matrix as carrier. (b) Morphology of tissue induction and regeneration in calvarial defects (light blue arrows) induced by gamma-irradiated osteogenic devices of 100 µg of hOP-1 combined with bovine insoluble collagenous bone matrix as carrier harvested on day 15 after surgical implantation (Ripamonti 2005). Pericranial islands of mineralized bone (dark blue arrow) induced as early as 15 days in adult non-human primates (Ripamonti 2005). (c, d) Prominent induction of bone formation by 100 (c) and 500 (d) µg of hOP-1 gamma-irradiated osteogenic devices harvested on day 30 after calvarial implantation showing rapid mineralization of the newly formed bone at the pericranial interface (dark blue arrows), with blocks of newly formed mineralized bone elevating the temporalis muscle (dark blue arrows), increasing the dose of the osteogenic device up to 500 µg

functions as a "frame" or "model" that shapes the body structure. This marvelous biological and molecular insight caused Aristotle to proffer a patterning function to the invading blood vessels, that is, "organogenetic blood vessels." The Aristotelian patterning scenario of sequential inductive and differentiating cascades of molecular and cellular events is cathartically condensed and summarized by the fascinating phenomenon of *"Bone: formation by autoinduction"* (Urist 1965; Reddi and Huggins 1972) and by the initiation of angiogenesis (Figure 2.5) upon implantation of naturally derived, highly purified BMPs in the non-human primate of the species *Papio ursinus* (Ripamonti et al. 1992).

Which are the molecular signals that set into motion the cascade of *"Bone: formation by autoinducton"* (Urist 1965; Reddi and Huggins 1972)? Several extracellular matrices of mammalian tissues, including uroepithelium, bone, and dentine, contain morphogenetic signals that initiate the induction of bone formation in heterotopic extraskeletal sites of a variety of animal models (Urist 1965; Reddi and Huggins 1972; Ripamonti et al. 2006; Ripamonti 2007). The uroepithelium, among other extracellular matrices, has been shown to possess the striking capacity to induce the heterotopic induction of bone formation, a phenomenon defined as "uroepithelial osteogenesis" (Huggins 1931; Huggins et al. 1936; Friedenstein 1961, 1968; Sacerdotti and Frattin 1901).

Osteogenesis is induced by transplantation of the urinary bladder, ligation of the renal artery, or surgical lesions of the wall of the urinary bladder (Friedenstein 1968). Of interest, the effects of transplantation of transitional epithelium differ significantly between mammals; in guinea pigs and feline models, transplantation of both auto- and allotransplants induces osteogenesis in a high percentage of cases (Friedenstein 1968). The osteogenic activity of transitional epithelium is highest in the guinea pig, feline, and canine models, lower in rodents, and lowest in lagomorphs (Friedenstein 1968).

---

**FIGURE 2.3**   *Continued*

(d) per gram of gamma-irradiated bovine inactive insoluble collagenous bone matrix as carrier (Ripamonti 2005). (e, f) Low-power microphotographs of undecalcified sections representing the whole calvarial defects regenerated after the implantation of 500 μg (e) and 2.5 mg (f) of gamma-irradiated hOP-1 combined with bovine insoluble collagenous bone matrix harvested on day 90 after implantation. Induction of solid blocks (dark blue arrows) of newly formed mineralized remodeling bone within the regenerated defects. Undecalcified blocks embedded in K-Plast resin with sections cut at 6 μm stained free-floating with Goldner's trichrome.

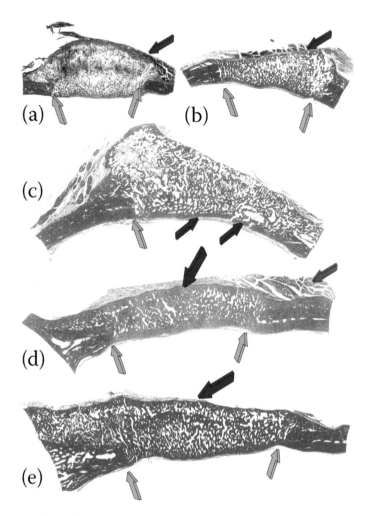

**FIGURE 2.4**  Effect of gamma irradiation of the hOP-1 osteogenic device on bone regeneration in calvarial defects of the Chacma baboon (*Papio ursinus*). Tissue induction and morphogenesis by non-gamma-irradiated recombinant human osteogenic protein-1 combined with allogeneic baboon or xenogeneic bovine insoluble collagenous bone matrices as carrier implanted in calvarial defects of adult non-human primate *P. ursinus* and harvested on days 30 (a), 90 (b, c) and 365 (d, e) after orthotopic calvarial implantation. (a) Prominent induction of pericranial (dark blue arrow) and endocranial bone across the entirety of the calvarial defect (light blue arrows), displacing the pericranium and associated temporalis muscle with a highly vascular connective tissue matrix between the pericranial and endocranial osteogenetic fronts with scattered insoluble collagenous matrix as carrier. (b) Prominent induction of bone formation extending the temporalis muscle pericranially (dark blue arrow) with the induction of solid remodeled osteonic bone endocranially across the defect (light blue arrows) on day 90. (c) Extensive induction of bone formation twofold higher than the normal calvaria 90 days after orthotopic calvarial implantation of the hOP-1 osteogenic device. (d, e) *Restitutio ad integrum* of the

Bladder transitional epithelium induces the differentiation of bone in allogeneic recipients, a phenomenon that Huggins has described as "the formation of bone under the influence of the epithelium of the urinary tract" or "uroepithelial osteogenesis" (Huggins 1931; Huggins et al. 1936). The transplanted allogeneic transitional epithelium of bladder mucosa in heterotopic intramuscular sites grows into solid cords of proliferating epithelium, around which bone forms concentrically, with osteoblasts facing the proliferating transitional epithelial cells. In his classic paper, Friedenstein (1968) dramatically illustrated cords of proliferating epithelium extending into the mesenchymal tissue surrounded by concentrically patterned newly formed bone with dividing osteoblasts in close proximity to the proliferating epithelium (Friedenstein 1968). Friedenstein (1968) superbly illustrates the fascinating scenario of uroepithelial osteogenesis (Huggins 1931; Huggins et al. 1936; Friedenstein 1961, 1962, 1968; Sacerdotti and Frattin 1901). Osteoblasts are seen proliferating and secreting bone matrix concentrically, surrounding solid cords of transitional epithelial cells grown into heterotopic sites of young pigs (Friedenstein 1968).

Huggins et al. (1936) conclude that the "proliferating mucosa of the kidney, ureter and bladder is a sufficiently strong stimulus to certain connective tissues in the dog and

---

**FIGURE 2.4** *Continued*

calvarial defects (light blue arrows) with prominent induction of bone formation greater than the recipient calvarium 365 days after implantation of non-gamma-irradiated bovine insoluble collagenous matrix as carrier for the recombinant morphogen (Ripamonti et al. 2000b, 2005). Other experiments, summarized in Figure 3.2, have shown that the limited tissue induction and bone morphogenesis, as seen in calvarial defects treated with gamma-irradiated osteogenic devices when compared to non-gamma-irradiated hOP-1 osteogenic devices (compare Figure 3.2 with Figure 4.2), are due not to the altered biological activity of gamma-irradiated hOP-1, but to the gamma irradiation damage of the insoluble collagenous bone matrix used as carrier (Ripamonti et al. 2000b, 2005; Ripamonti 2005). Northern blot analyses of tissue specimens harvested from calvarial sites implanted with doses of the gamma-irradiated osteogenic devices showed prominent induction of OP-1 mRNAs throughout the time periods of 15, 30, 90, and 365 days after implantation, together with a biphasic spatial and temporal expression of TGF-$\beta_1$, which has indicated a specific temporal transcriptional window during the induction of bone formation in primates. This has been further highlighted by the demonstration of the induction of bone formation in heterotopic intramuscular sites of *P. ursinus* by the three mammalian TGF-$\beta$ isoforms (Ripamonti et al. 1997, 2000b, 2008, 2014; Ripamonti and Roden 2010; Klar et al. 2014); TGF-$\beta$ isoforms induce bone formation by initiating the expression of several BMPs genes upon the implantation of a mammalian TGF-$\beta$ isoform (Ripamonti et al. 2014; Klar et al. 2014). Undecalcified blocks embedded in K-Plast resin with sections cut at 6 µm stained free-floating with Goldner's trichrome.

rabbit to induce the formation of bone." Huggins further noted that it is the proliferating mucosa of the renal pelvis, ureter, and bladder that have the capacity of inducing osteogenesis, and not the transplanted nonproliferating epithelial cells per se. Indeed, Huggins (Huggins 1931; Huggins et al. 1936) concluded that "the proliferating newly formed epithelium and not the non-proliferating part of the transplant is the essential factor in this osteogenesis," that is, uroepithelial osteogenesis.

An important experiment resulting in the induction of bone formation with hematopoietic bone marrow in the kidney was reported after ligation of the renal vascular pedicle in the rabbit (Sacerdotti and Frattin 1901). It was shown that the kidney parenchyma was transformed into bone with trabeculation and the associated induction of hematopoietic bone marrow. On day 90 after ligation, Sacerdotti and Frattin (1901) observed the generation of true bone with hematopoietic marrow, suggesting that the induction of bone resulted from metaplastic changes of the underlying connective tissue stroma, ultimately inducing membranous ossification, as observed in craniofacial bones. Sacerdotti and Frattin (1901) further elaborated that the ligation of the renal artery in rabbits shows unequivocally that it is possible to *de novo* generate bone with bone marrow within tissues that normally do not contain osteogenic tissue, that is, to induce *de novo* heterotopic bone formation (Sacerdotti and Frattin 1901).

Friedenstein (1962) ultimately asked the compelling question that perforce defined the bone induction principle (Urist et al. 1967, 1968) or the osteogenic activity of several transplanted tissues, including bone and dentine matrices, and uroepithelium (Urist et al. 1967, 1968; Levander and Willestaedt 1958; Urist and McLean 1952): How is the inductive influence of the transitional epithelium transferred to the competent responding cells? Friedenstein elegantly hypothesized the humoral nature of the osteogenic activity of transitional epithelium, that is, the presence of a soluble molecular signal or "inductor" (Friedenstein 1962).

Last century's research has been perforce speculative until the incisive work of Wozney et al. (1988), Celeste et al. (1990), and Özkaynak et al. (1990) cloned a new class of gene products, the bone morphogenetic proteins (BMPs), which are members of the transforming growth factor-$\beta$ (TGF-$\beta$) supergene family. Importantly, Özkaynak et al., in further work, reported the molecular link to the uroepithelial osteogenesis, showing high levels of osteogenic protein-1 (OP-1) mRNA in the kidney (Özkaynak et al. 1991).

The inductive potency of the bladder mucosa to induce uro-epithelial osteogenesis has also been tested by transplanting the dome of the bladder into the *rectus abdominis* muscle fascia and vice versa by transplanting the fascia of the *rectus abdominis* into full thickness defects of the dome of the bladder of the non-human primate *P. ursinus* (Ripamonti, Bone Research Laboratory, unpublished data; Ripamonti et al. 2002).

Friedenstein (1968), reviewing the transplantation of transitional epithelial cells rather than whole segments of bladder mucosa as reported by Huggins (Huggins 1931; Huggins et al. 1936), asked: "How is the inducible influence of transitional epithelium transferred to the competent cells?" Friedenstein postulated the presence of an "inductor," that is, a substance produced by epithelial cells that would induce phenotypic and genotypic changes in adjacent responding cells. Importantly, Friedenstein made the key observation that only the epithelium lining the basement membrane possesses osteogenic properties. Indeed, the epithelium, when detached from the tunica propria by trypsinization, induces osteogenesis, while the remaining tunica propria lacks inductive properties (Friedenstein 1968). Friedenstein also observed transfilter bone formation by transitional epithelium with foci of newly formed bone on the outer surface of the Millipore filter, which naturally implied the existence of a soluble inductor, that is, a diffusible molecular signal or morphogen (Friedenstein 1962, 1968).

Levander (1938), in his classic paper "A Study on Bone Regeneration," briefly summarized how connective tissue "may be transformed into bone tissue." In his search for "a substance with bone forming properties" in bone grafts, including alcoholic extracts of bone matrix as well as reparative calluses (Levander and Willestaedt 1958), Levander (1938) concluded that heterotopic formation of bone is induced by "some substance extracted by alcohol from the skeletal tissue, a substance having the power to activate the non-specific mesenchymal tissue into the formation of bone tissue, either directly or *via* the embryonic prenatal stage of bone, viz., cartilage." Levander is thus credited to have been among the first who has stated that the induction of bone formation in postnatal life recapitulates events that occur in the normal course of embryonic development (Levander 1938). In his experiments, Levander (1938) always observed that the implanted tissue is very rich in vessels and responding inducible cells. Cells group around invading vessels and large cells in perivascular locations are seen with hyperchromatic nuclei. Levander (1938) further hypothesized that "the impression is given by these pictures that the fully formed mesenchymal cells ultimately emanate from the endo-thelial cells of the sprouting capillaries."

Levander (1938) and Trueta (1963) ascribed to the invading vessels not only osteogenetic but also morphogenetic preroga-tives, as stated by Aristotle (Lanza and Vegetti 1971; Crivellato et al. 2007). Levander (1945) described the tissue induced by alcoholic extracts of bone matrices as newly engineered tissue characterized by prominent osteogenesis and capillary sprout-ing surrounded by mesenchymal condensations with abundant perivascular cells, also introducing the term *tissue induction*. Both Levander (1938, 1945) and Trueta (1963) suggested that perivascular hyperchromatic stem cells contribute to bone depo-sition around the vessels, which are thus both osteogenetic and morphogenetic as per Aristotelian insights (Lanza and Vegetti 1971; Crivellato et al. 2007).

Morphogenesis, the genesis of form and function (Reddi 1984; Lander 2007; Kerszberg and Wolpert 2007), includes pattern formation with three-dimensional parameters con-structing the architecture of functional tissue. Angiogenetic and osteogenetic vessels mold the newly formed bone by induc-tion on and around the sprouting and proliferating capillaries (Levander 1938, 1945; Trueta 1963). The three-dimensional construct is thus formed around the sprouting capillaries, which, together with the axial three-dimensional pattern of tissue growth, bring about endothelial and pericytic responding stem cells in the perivascular location. These, together with the basement membrane of the sprouting capillaries, a prominent and rich source of molecular signals initiating, regulating, and orchestrating both angiogenesis and osteogenesis, ultimately control and define *"Bone: Formation by autoinducton"* (Urist 1965; Reddi and Huggins 1972). The invading three-dimensional pattern of vasculogenesis with capillary sprouting and invasion constructing the primate corticocancellous bone is depicted in a series of unique digital images shown in Figure 2.5.

Angiogenesis with capillary sprouting, growth, and differenti-ation is the three-dimensional construct for the induction of bone formation (Figure 2.5). Each central blood vessel is surrounded by mesenchymal cellular condensations that must include peri-vascular stem cells, differentiating pericytes, and other stem cell progenitors in close contact with the capillary basement mem-brane. Cellular condensations around each morphogenetic and osteogenetic vessel construct the three-dimensional architecture of the primate cortico-cancellous Haversian bone (Ripamonti 2006; Ripamonti et al. 2006) and its *quantum* structure, the osteosome (Reddi 1997).

Intuitively, the high-power digital images presented in Figure 2.5c–e indicate that the sprouting morphogenetic capillary secretes soluble molecular signals that induce pattern formation and morphogenesis. The vessels thus construct a series of highly repetitive geometric constructs each modeled as osteosomes, the *quantum* anatomy of the lamellar osteonic bone (Parfitt 1994; Manolagas and Jilka 1995; Parfitt et al. 1996; Reddi 1997). Perivascular niches contain several different stem cells (Benjamin et al. 1998; Kovacic and Boehm 2009), recently identified in striated primate muscle as myoendothelial cells (Zheng et al. 2007), possibly a source of several molecular signals controlling tissue induction and remodeling of the surrounding extracellular matrix. Endothelial cell-derived microparticles (MPs) are small membrane vesicles associated with matrix remodeling (Lozito and Tuan 2011) that may secrete a battery of matrix molecules, including mRNAs controlling matrix synthesis and remodeling at a distance from the secreting endothelial cells.

Capillary sprouting during tissue induction and morphogenesis with the induction of cellular condensations surrounding each morphogenetic and osteogenetic vessel is the prototype example of tissue induction and morphogenesis of bone, as shown in Figure 2.5. This is initiated by the biomimetism of the extracellular matrix, which includes soluble and insoluble signals that construct the complex three-dimensional architecture of the bone/bone marrow organ. Figure 2.5 shows that tissue induction and morphogenesis initiate around an expanding three-dimensional tissue construct of sprouting capillaries. The invading capillaries *per se* contain all the necessary molecular signals to induce tissue induction and morphogenesis, that is, soluble, insoluble, and biomimetic extracellular matrix signals, including morphogenetic sequences of laminin and type IV collagen within the basement membrane of the invading capillaries (Vukicevic et al. 1990), whose motifs are read across the basement membrane by the endothelial/osteoblastic microenvironments, setting into motion the ripple-like cascade of cell differentiation and tissue induction (Vukicevic et al. 1990).

The complex cellular, molecular, and mechanical signals that regulate the assembly of the extracellular matrix precisely regulate angiogenesis and vascular invasion (Reddi 1994, 2000; Ingber and Folkman 1989; Ingber et al. 2006; Crivellato et al. 2007). Capillary sprouting of the osteogenetic vessels within the highly vascular mesenchyme is followed by condensations of angioblastic and mesenchymal primitive tissue around each osteogenetic vessel within highly cellular mesenchymal

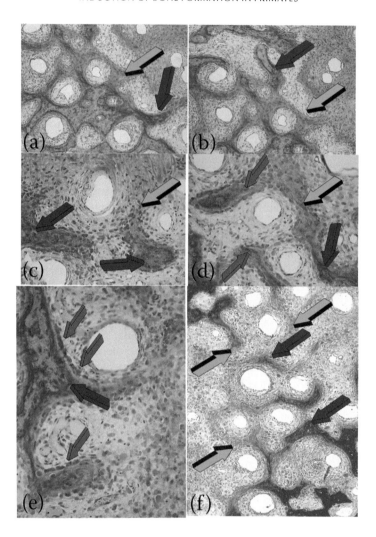

**FIGURE 2.5** Angiogenesis and the induction of bone formation by morphogenetic and osteogenetic vessels of Aristotle's (Crivellato et al. 2007) and Trueta's (1963) definitions. Vascular invasion and capillary sprouting provide the molecular, cellular, and morphological templates for the induction of bone formation by highly purified, naturally derived osteogenic fractions implanted in the non-human primate *Papio ursinus*. Single invading capillaries, the osteogenetic vessels of Trueta's (1963) description and definition, dictate the pattern of the induction of bone formation and act as a template for the induction of the Haversian primate osteonic bone. The invading vessels are also morphogenetic since they initiate the induction of mesenchymal cellular condensations around each invading capillary. (a) Detail of invading osteogenetic vessels constructing cellular differentiation and mesenchymal condensations (light blue arrow) with foci of early-onset mineralization (dark blue arrow) developing around each single capillary that controls the differentiation and development of lamellar osteonic bone. (b, c) Newly developed mesenchymal cellular condensations embrace

condensations rich in angioblastic and perivascular stem cells (Ripamonti 2007) (Figure 2.5). The invading capillaries are engineering the architecture of the newly forming bone. Bone forms by the organization of mesenchymal condensations around each patterning capillary; the capillary is thus morphogenetic, constructing the pattern of the induction of bone formation (Figure 2.5). Perivascular and vascular condensations of mesenchymal and angioblastic origin pattern the Haversian canal system of the primate osteonic bone, whereby each central blood vessel is surrounded by newly formed bone with nascent foci of mineralization surfaced by osteoid seams populated by differentiated contiguous osteoblasts (Figure 2.5).

The three-dimensional construct of the invading mesenchymal condensations thus provides the structural framework for the differentiation of osteoblastic cells, osteoblastic synthesis, matrix deposition, and mineralization of the collagenous matrix, with foci of mineralization within the collagenous condensations embedding osteocytes within the newly formed and mineralized bone (Figure 2.5e and f). Patterning and mineralizing mesenchymal condensations surfaced by osteoblastic cells provide the structural framework for the exquisite intimate relationship between the endothelial/pericytic cells of the osteogenetic vessels and the osteoblastic cells facing the morphogenetic and osteogenetic vessels. Capillary sprouting and invasion are prerequisite for osteogenesis since both angiogenic and osteogenic proteins are bound to type IV collagen of the basement membrane of the invading capillaries (Paralkar et al. 1990, 1991; Folkman et al. 1988). Importantly, the binding and sequestration of both angiogenic and osteogenic proteins to the basement membrane components of the patterning capillaries provide the conceptual framework of the supramolecular assembly of the newly formed vascularized bone (Paralkar et al. 1990, 1991; Ripamonti and Reddi 1992; Folkman et al. 1988; Ripamonti et al. 2006). Basement membrane components, by sequestering both initiators and promoters of angiogenesis and osteogenesis (Paralkar et al. 1990, 1991; Vlodavsky et al. 1997; Folkman et al. 1988), are directly modeling bone formation in angiogenesis (Figure 2.5).

**FIGURE 2.5** *Continued*

each invading capillary differentiating osteoblastic-like cells (light blue arrows) facing the central osteogenetic vessels, now controlling the genesis of the corticocancellous osteonic primate bone. Mesenchymal condensations mineralize (dark blue arrows), thus patterning the osteonic bone covered by plump osteoblast-like cells facing the central morphogenetic and osteogenetic vessels (b–e).

Angiogenesis is a prerequisite for osteogenesis (Trueta 1963). Osteoprogenitors and osteoblasts are in contact with the basement membrane of the invading capillaries. The endothelial cell matrix may function as a morphogenetic cue during the critical phase of capillary invasion (Paralkar et al. 1990, 1991; Ripamonti 2006).

*In vitro* morphological and molecular studies have suggested that bone-forming cells in contact with the basement membrane of the invading capillaries set into motion the formation of a matrix/cytoplasmic reticulum resembling the osteocyte's canalicular network of the bone matrix. Of note, Reddi's team suggested that the osteocyte, a developmental stage of the osteprogenitor–osteoblast lineage, may retain a memory of the initial contact of the osteoblast with extracellular matrix components and selected amino acid motifs of the invading capillaries, laminin and type IV collagen, when reading amino acid sequence motifs across the membrane extracellular matrix (Vukicevic et al. 1990). This initial contact may set into motion the ripple-like cascade of cell differentiation and the induction of bone formation (Vukicevic et al. 1990).

## 2.3  Homologous but Molecularly Different Pleiotropic Proteins of the TGF-β Supergene Family Initiate Endochondral Bone Formation

The induction of bone formation during embryonic development and postnatal tissue induction and morphogenesis is set by complex, seemingly redundant cascades of molecular and morphogenetic events that sculpt the multicellular mineralized and nonmineralized structures of the bone/bone marrow organ, ultimately engineering the emergence of the skeleton and thus of the vertebrates (Reddi 1997, 2000; Ripamonti 2006, 2007, 2009).

The transforming growth factor-β (TGF-β) supergene family comprises several multifactorial gene products endowed with a vast pleiotropic activity regulating embryonic development, tissue patterning, postnatal tissue induction and morphogenesis, immunoregulation, and fibrosis, initiating the morphogenesis of several tissues and organs, including the skeleton and the bone matrix (Reddi 1997, 2000; Wozney et al. 1988; ten Dijke et al. 1990; Massagué 2000; Ripamonti 2006, 2009).

Recent studies have indicated additional specific roles of the three mammalian TGF-β isoforms within the bone matrix

(Balooch et al. 2005; Spagnoli et al. 2007). Balooch et al. (2005) identified TGF-βs as the key regulators of the mechanical properties and composition of the bone matrix, and additionally as the ultimate morphogen to maintain functional parameters of bone quality, bone mass, elastic modulus and hardness, mineral concentration, and resistance to fracture, controlling the structure and function of the skeleton. Recent studies have also shown that mRNA expression of TGF-β gene products regulates joint morphogenesis (Spagnoli et al. 2007).

Several preclinical studies in a variety of animal models, including non-human primates, have taught us that the BMPs are a family of highly conserved secreted pleiotropic proteins that initiate endochondral bone differentiation *in vivo* (Figures 2.1 through 2.4) (Reddi 2000; Ripamonti et al. 1992, 2000, 2005; Ripamonti 2006b).

The research work of Sampath et al. (1993) in the fruit fly *Drosophila melanogaster* showed that there is a high level of homology between *Decapentaplegic* (*dpp*) and *60A* genes in *D. melanogaster* with human BMP-2, BMP-4, and BMP-5, and BMP-6, respectively (Sampath et al. 1993). The study thus indicated the primordial role of BMP sequence motifs during the emergence of the vertebrates (Ripamonti 2003).

Because of evolutionary and functional conservation, the secreted proteins have retained common developmental roles. Indeed, the most compelling evidence that gene products in the fruit fly *D. melanogaster* and *H. sapiens* have been conserved for almost a billion years is that recombinant human DPP and 60A proteins induce heterotopic endochondral bone differentiation in the rodent subcutaneous assay (Sampath et al. 1993).

Grandly, the induction of bone formation by *dpp* and *60A* gene products in the mammalian heterotopic bioassay (Sampath et al. 1993) has shown that gene products of the fruit fly *D. melanogaster* and *H. sapiens* are conserved through a billion years of molecular evolution, and that a phylogenetically ancient signaling carboxy-terminal domain deployed for dorsoventral patterning in the fruit fly *D. melanogaster* is additionally operational to construct the unique vertebrate trait, that is, the induction of bone formation; skeletogenesis and the emergence of the skeleton; the vertebrate mammals; the emergence of the ancient bipedal hominids, the Australopithecinae, early *Homo* species; and at last, the explosion of the Homo clade (Ripamonti 2007, 2009).

A series of systematic studies in the non-human primate *P. ursinus* provided evidence for a novel function of the mammalian TGF-β isoforms, that is, the induction of bone formation in heterotopic extraskeletal sites. The three mammalian recombinant human TGF-β isoforms, when reconstituted with either allogeneic insoluble collagenous bone matrix or coral-derived hydrothermally exchanged hydroxyapatite/calcium carbonate macroporous bioreactors, initiate the rapid and substantial induction of bone formation in extraskeletal heterotopic sites of *P. ursinus* (Ripamonti et al. 1997, 2000a, 2008, 2012, 2014a; Ripamonti and Roden 2010; Klar et al. 2014). Reconstituted bioreactors superactivated by the hTGF-β isoforms (Ripamonti et al. 2012) were implanted heterotopically in the *rectus abdominis* muscle of *P. ursinus* (Duneas et al. 1998; Ripamonti et al. 1997, 2000a, 2008, 2012, 2014; Ripamonti and Roden 2010; Klar et al. 2014), additionally providing the scientific basis for synergistic molecular therapeutics for the rapid induction of endochondral bone in primate species (Figure 2.6) (Ripamonti et al. 1997, 2010; Duneas et al. 1998).

The induction of endochondral bone formation by the recombinant hTGF-$\beta_1$ in heterotopic intramuscular sites in the primate (Figure 2.6a) (Ripamonti et al. 1997) was unexpected in the light of previously established results in rodents that showed the lack of bone induction by the hTGF-β isoforms in the heterotopic bioassay in rodents (Roberts et al. 1986; Sampath et al. 1987; Hammonds et al. 1991; Shinozaki et al. 1997; Shah et al. 1995).

In a first set of experiments in the *rectus abdominis* muscle of *P. ursinus*, 5 μg of hTGF-$\beta_1$ formed mineralized corticalized ossicles by day 30 (Figure 2.6a). Binary applications of recombinant hOP-1 and hTGF-$\beta_1$ at a ratio of 20:1 hOP-1/hTGF-$\beta_1$ resulted in the rapid induction of large corticalized ossicles by day 15 (Figure 2.6b), with newly formed bone trabeculae covered by osteoid seams (Figure 2.6c). Importantly, the induction of bone formation resulted in the morphogenesis of structurally organized cartilage zones, highly reminiscent of embryonic growth plates (Figure 2.6d). The digital image represented in Figure 2.6d strongly relates to a mammalian developmental growth plate and vividly illustrates the concept that regeneration of cartilage and bone in postnatal life shares common cellular and molecular mechanisms with embryonic developments, and that the "memory" of developmental events in the embryo can be redeployed postnatally by the application of morphogen combinations (Ripamonti et al. 1997).

Chapter 6, which presents Figure 6.2, describes the synergistic induction of bone formation as observed in seminal studies in *P. ursinus* (Ripamonti et al. 1997, 2010; Duneas et al. 1998). The induction of bone formation in both heterotopic and heterotopic intramuscular sites is raised several-fold when compared to the induction of bone formation by doses of a single recombinant hBMP: binary applications of hOP-1 with relatively low doses of hTGF-$\beta_1$ or naturally derived porcine TGF-$\beta_1$ (pTGF-$\beta_1$) at a ratio of 20:1 by weight raise several-fold the osteogenic activity of the implanted synergistic devices in heterotopic and orthotopic calvarial sites (Figure 2.6c–d).

Our laboratories have studied extensively the induction of bone formation by the mammalian hTGF-$\beta$ isoforms and hypothesized that TGF-$\beta$ signaling induces endochondral bone differentiation by regulating Noggin expression (Ripamonti and Roden 2010), and therefore BMP activity (Groppe et al. 2003; Gazzerro et al. 1998). We have further postulated that if these molecular and cellular scenarios are correct, addition of recombinant human Noggin (hNoggin) together with a mammalian TGF-$\beta$ isoform would inhibit the osteogenic activity of expressed and secreted bone morphogenetic proteins, resulting in limited bone formation by induction (Ripamonti and Roden 2010; Ripamonti et al. 2010). Our previous studies and combined data using the three mammalian recombinant human TGF-$\beta$ isoforms have shown that the TGF-$\beta$ proteins may act upstream of the bone morphogenetic proteins and may induce the induction of endochondral bone formation by expressing BMP-related gene products, ultimately directing the induction of bone formation (Duneas et al. 1998; Ripamonti et al. 2000a, 2008; Ripamonti and Roden 2010).

We have recently shown that hTGF-$\beta_3$ signaling induces endochondral bone differentiation by regulating Noggin expression and BMP/OP activity, resulting in the induction of bone formation (Ripamonti et al. 2014; Klar et al. 2014). Binary applications of 125 μg of hNoggin with equal doses of hTGF-$\beta_3$ significantly inhibit the extent of bone differentiation when compared to coral-derived macroporous bioreactors superactivated by the hTGF-$\beta_3$ isoform (Ripamonti et al. 2014; Klar et al. 2014).

Chapter 3 extensively describes the bone inductive activity of hTGF-$\beta_3$ when implanted in heterotopic *rectus abdominis* sites of *P. ursinus* as a prerequisite for potential clinical applications in humans. The results so far achieved in non-human and human primates using the hTGF-$\beta_3$ osteogenic device are challenging the *status quo* of the bone tissue engineering paradigm, and paraphrasing Collins's (2011) title in *Science Translational Medicine*, the time is now right to reengineer the induction of bone formation by translating the hTGF-$\beta_3$ osteogenic device in clinical contexts, and to reevaluate the induction of bone formation in primate models, including humans (Ripamonti et al. 2014).

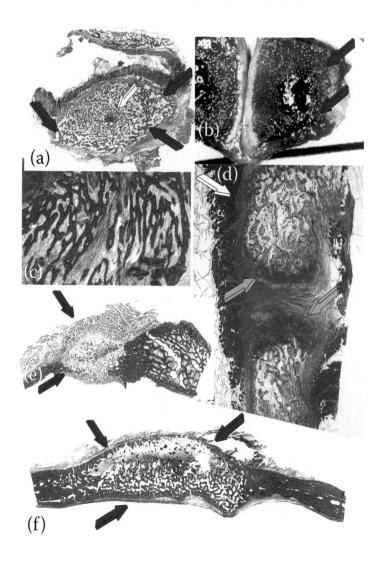

**FIGURE 2.6** Redundancy of soluble molecular signals initiating the induction of bone formation in heterotopic intramuscular sites of the Chacma baboon (*Papio ursinus*). (a) Induction of a large corticalized ossicle upon the implantation of 5 μg of recombinant human transforming growth factor-β₁ (hTGF-β₁) reconstituted with 100 mg of allogeneic insoluble collagenous bone matrix implanted in the rectus abdominis muscle of *P. ursinus* and harvested on day 30 (Ripamonti et al. 1997). Corticalization of the newly formed ossicle with mineralized newly formed bone (dark blue arrows) surrounds scattered remnants of insoluble collagenous matrix as carrier (white arrow). (b) Complete morphogenesis of a large ossicle by day 15 upon implantation of synergistic binary application of hOP-1/hTGF-β₁ 20:1 by weight (Ripamonti et al. 1997). (c) Undecalcified section of the 15-day heterotopic ossicle characterized by trabeculation of mineralized newly formed bone in blue surfaced by osteoid seams populated by contiguous osteoblasts.

## Acknowledgments

The noncanonical induction of bone formation by the mammalian transforming growth factor-β isoforms has been the source of fascinating intellectual stimulation, particularly for the vision of changing the paradigm of bone tissue engineering and skeletal reconstruction in clinical contexts. I would like to thank the University of the Witwatersrand, Johannesburg, the National Research Foundation, and *ad hoc* grants to the Bone Research Laboratory for their support of the continuous and systematic studies in non-human primates since the early 1990s on the induction of bone formation by the mammalian transforming growth factor-β isoforms. I thank the several students, laboratory technologists, scientists, and visiting professors who have significantly contributed to the several discoveries highlighted in this book, in particular Barbara van den Heever, Laura Yeates, Ruqayya Parak, Manolis Heliotis, Carlo Ferretti, Roland Klar, Therese Dix-Peek, Brenda Milner, Raquel Duarte, and Peter ten Dijke. A special word of thanks to M.R. Urist, A.H. Reddi, and *"Bone: Formation by Autoinduction"* for continuously providing the inspiration to create, discover, and write.

---

**FIGURE 2.6** *Continued*

(d) Tissue induction and morphogenesis after implantation of interposed recombinant hOP-1 and hTGF-β₁ separated by the linea alba of the *rectus abdominis* muscle (white arrow) in *P. ursinus*. Maturational gradients of tissue induction and morphogenesis at the periphery of newly formed ossicles after the synergistic induction of bone formation by the implanted interposed recombinant morphogens (Ripamonti et al. 1997). Cartilage differentiation (light blue arrows) surfaces trabeculae of newly formed bone at the site of two juxtaposed ossicles. There is a gradient of morphological structures that engineers in heterotopic sites a rudimentary growth plate additionally characterized by the patterned palisading of chondroblastic/chondrocytic cells aligned on the surface of the newly formed bone. (e, f) Substantial rapid induction of bone formation by binary application of hOP-1 (100 µg) and porcine platelet-derived transforming growth factor-β₁ (pTGF-β₁) in calvarial defects of *P. ursinus* harvested on days 30 (e) and 90 (f) after calvarial implantation. Extensive pericranial bone deposition (dark blue arrows) surrounding central scattered remnants of collagenous matrix as carrier. The induction of bone formation displaces the temporalis muscle. Undecalcified blocks embedded in K-Plast resin with undecalcified sections cut at 6 µm stained free-floating with Goldner's trichrome.

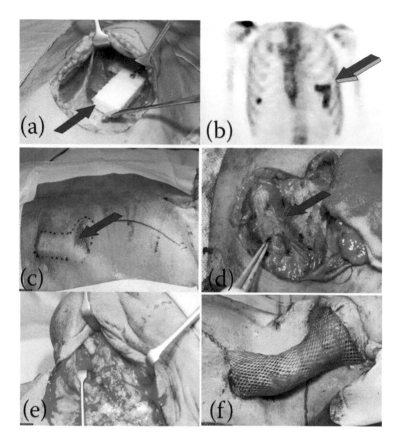

**FIGURE 2.7** Construction of a prefabricated human osteogenic flap in heterotopic intramuscular sites for later transplantation into a mandibular recipient bed for mandibular reconstruction after squamous cell carcinoma debridement and surgical ablation of a large mandibular segment (Heliotis et al. 2006). The privileged highly vascularized intramuscular sites with niches of responding perivascular and vascular stem cells (reviewed in Ripamonti et al. 2007; Ripamonti 2009; Kovacic and Boehm 2009) have generated the concept of manufacturing prefabricated heterotopic bone grafts for autologous transplantation (reviewed in Ripamonti et al. 2006; Ripamonti et al. 2007). (a) An L-shaped coral-derived macroporous construct (dark blue arrow) is implanted in the muscular tissue of the chest of a human patient after reconstitution with 2.5 mg doses of the osteogenic protein-1 device. (b) Scintigraphy demonstrates the induction of bone formation (dark blue arrow) in the L-shaped prefabricated flap in the pectoralis muscle. (c–e) Surgical debridement and harvest of the newly generated L-shaped sagomated heterotopic prefabricated graft and autologous transplantation into the surgically prepared recipient bed (c–e) after the preparation of a pedicled flap (c, f) into the mandibular recipient bed (Heliotis et al. 2006).

# References

BALOOCH, G., BALOOCH, M., NALLA, R.K., SCHILLING, S., FILVAROFF, E.H., MARSHALL, G.W., MARSHALL, S.J., RITCHIE, R.O., DERYNCK, R., ALLISTON, T. (2005). TGF-beta regulates the mechanical properties and composition of bone matrix. *Proc Natl Acad Sci USA* 102(52), 18813–18.

BENJAMIN, L.E., HOMO, I., KESHET, E. (1998). A plasticity window for blood vessel remodelling is defined by pericyte coverage of the preformed endothelial network and is regulated by PDGF-B and VEGF. *Development* 125, 1591–98.

BYRNE, A.M., BOUCHIER-HAYES D.J., HARMAY, J.H. (2005). Angiogenic and cell survival functions of vascular endothelial growth factor. *J Cell Mol Med* 9, 777–94.

CARLEVARO, M.F., CERMELLI, S., CANCEDDA, R., DESCALZI, CANCEDDA, F. (2000). Vascular endothelial growth factor (VEGF) in cartilage neovascularization and chondrocyte differentiation: Autoparacrine role during endochondral bone formation. *J Cell Sci* 113(Pt 1), 59–69.

CELESTE, A.J., IANNAZZI, J.M., TAYLOR, J.A., HEWICK, R.C., ROSEN, V., WANG, E.A., WOZNEY, J.M. (1990). Identification of transforming growth factor-$\beta$ family members present in bone-inductive protein purified from bovine bone. *Proc Natl Acad Sci USA* 87, 9843–47.

COLLINS, F.S. (2011). Reengineering translational science: The time is right. *Sci Transl Med* 3(90), 90cm17. doi: 10.1126/scitranslmed.3002747.

CRIVELLATO E., NICO B, RIBATTI D. (2007). Contribution of endothelial cells to organogenesis: A modern reappraisal of an old aristotelian concept. *J Anat* 211, 415–27.

DAL MAS, I., BISCARDI, A., SCHNITZLER, C.M. (2007). Bone loss in the ovariectomized baboon *Papio ursinus*: Densitometry, histomorphometry and biochemistry. *J Cell Mol Med* 11(4), 852–67.

DE ROBERTIS, E.M. (2008). Evo-devo: Variations on ancestral themes. *Cell* 132, 185–95.

DUNEAS, N., CROOKS, J., RIPAMONTI, U. (1998). Transforming growth factor-$\beta_1$: Induction of bone morphogenetic protein genes expression during endochondral bone formation in the baboon, and synergistic interaction with osteogenic protein-1 (BMP-7). *Growth Factors* 15, 259–77.

FOLKMAN, J., D'AMORE, P.A. (1996). Blood vessel formation: What is its molecular basis? *Cell* 87, 1153–55.

FOLKMAN, J., KLAGSBRUN, M., SASSE, J., WADZINSKI, M., INGBER, D., VLODAVSKY, I. (1988). A heparin-binding angiogenic protein—basic fibroblast growth factor—is stored within basement membrane. *Am J Pathol* 130(2), 393–400.

FRIEDENSTEIN, A.J. (1961). Osteogenic activity of transplanted transitional epithelium. *Acta Anat* (Basel) 45, 31–59.

FRIEDENSTEIN, A.J. (1962). Humoral nature of osteogenic activity of transitional epithelium. *Nature* 194, 698–99.

FRIEDENSTEIN, A.J. (1968). Induction of bone tissue by transitional epithelium. *Clin OrtopRel Res* 59, 21–37.

GAZZERRO, E., GANGJI, V., CANALIS, E. (1998). Bone morphogenetic proteins induce the expression of noggin, which limits their activity in cultured rat osteoblasts. *J Clin Invest* 102(12), 2106–14.

GERBER, H.P., VU, T.H., RYAN, A.M., KOWALSKI, J., WERB, Z., FERRARA, N. (1999). VEGF couples hypertrophic cartilage remodeling, ossification and angiogenesis during endochondral bone formation. *Nat Med* 5, 623–28.

GROPPE, J., GREENWALD, J., WIATER, E., RODRIGUEZ-LEON, J., ECONOMIDES, A.N., KWIATKOWSKI, W., BABAN, K., AFFOLTER, M., VALE, W.W., IZPISUA BELMONTE, J.C., CHOE, S. (2003). Structural basis of BMP signaling inhibition by Noggin, a novel twelve-membered cystine knot protein. *J Bone Joint Surg Am* 85A(Suppl 3), 52–58.

HAMMONDS, R.G., JR., SCHWALL, R., DUDLEY, A., BERKEMEIER, L., LAI, C., LEE, J., CUNNINGHAM, N., REDDI, A.H., WOOD, W.I., MASON, A.J. (1991). Bone-inducing activity of mature BMP-2b produced from a hybrid BMP-2a/2b precursor. *Mol Endocrinol* 5(1), 149–55.

HELIOTIS, M., LAVERY, K.M., RIPAMONTI, U., TSIRIDIS, E., DI SILVIO, L. (2006). Transformation of a prefabricated hydroxyapatite/osteogenic protein-1 implant into a vascularised pedicled bone flap in the human chest. *Int J Oral Maxillofacial Surg* 35, 265–69.

HUGGINS, C.B. (1931). The formation of bone under the influence of epitheliam of the urinary tract. *Arch Surg* 22, 377–408.

HUGGINS, C.B., McCARROL, H.R., BLOCKSOM, B.H. (1936). Experiments on the theory of osteogenesis: The influence of local calcium deposits on ossification; the osteogenic stimulus of epithelium. *Arch Surg* 32, 915–31.

INGBER, D.E., FOLKMAN, J. (1989). How does extracellular matrix control capillary morphogenesis? *Cell* 58(8), 803–5.

INGBER, D.E., MOW, V.C., BUTLER, D., NIKLASON, L., HUARD, J., MAO, J., YANNAS, I., KAPLAN, D., VUNJAK-NOVAKOVIC, G. (2006). Tissue engineering and developmental biology: going biomimetic. *Tissue Eng* 12(12), 3265–83.

KEITH, A. (1927). Concerning the origin and nature of osteoblasts. *Proc R Soc Med* 21, 301–8.

KERSZBERG, M., WOLPERT L. (2007). Specifying positional information in the embryo: Looking beyond morphogens. *Cell* 130, 205–9.

KHOURI, R.K., KOUDSI, B., REDDI, H. (1991). Tissue transformation into bone in vivo: A potential practical application. *JAMA* 266(14), 1953–55.

KLAR, R.M., DUARTE, R., DIX-PEEK, T., RIPAMONTI, U. (2014). The induction of bone formation by the recombinant human transforming growth factor-$\beta_3$. *Biomaterials* 35(9), 2773–88.

KOVACIC, J.C., BOEHM, M. (2008). Resident vascular progenitor cells: An emerging role for non-terminally differentiated vessel-resident cells in vascular biology. *Stem Cell Res* 2, 2–15.

LANDER, A.D. (2007). Morpheus unbound: Reimagining the morphogen gradient. *Cell* 128, 245–56.

LANZA D., VEGETTI M. (1971). *Opere biologiche: Di Aristotele. A cura di Diego Lanza e Mario Vegetti.* Torino: Utet.

LEUNG, D.W., CACINES, G., KUANG W.J., GOEDDEL, D.W., FERRARA, N. (1989). Vascular endothelial growth factor is a secreted angiogenic mitogen. *Science* 246, 1306–9.

LEVANDER, G. (1938). A study on bone regeneration. *Surg Obstetr* 67, 705–14.

LEVANDER, G. (1945). Tissue induction. *Nature* 155, 148–49.

LEVANDER, G., WILLESTAEDT, H. (1946). Alcohol-soluble osteogenetic substance from bone marrow. *Nature* 3992, 587.

LOZITO, T.P., TUAN, R.S. (2011). Mesenchymal stem cells inhibit both endogenous and exogenous MMPs via secreted TIPMs. *J Cell Physiol* 226(2), 385–96.

MANOLAGAS, S.C., JILKA, R.L. (1995). Bone marrow, cytokines, and bone remodeling. Emerging insights into the pathophysiology of osteoporosis. *N Engl J Med* 332(5), 305–11.

MASSAGUÉ, J. (2000). How cells read TGF-beta signals. *Nat Rev Mol Cell Biol* 1(3), 169–78.

MOSS, M.L. (1958). Extraction of an osteogenic inductor factor from bone. *Science* 127, 755–56.

ÖZKAYNAK, E., RUEGER, D.C., DRIER, E.A., CORBETT, C., RIDGE, R.J., SAMPATH, T.K, OPPERMANN, H. (1990). OP-1 cDNA encodes anosteogenic protein in the TGF-beta family. *EMBO J* 9(7), 2085–93.

ÖZKAYNAK, E., SCHNEGELSBERG, P.N. OPPERMANN, H. (1991). Murine osteogenic protein-1 (OP-1): High levels of mRNA in kidney. *Biochem Biophys Res Commun* 179(1), 116–23.

PARALKAR, V.M., NANDEDKAR, A.K.N., POINTER, R.H., KLEINMAN, H.K., REDDI, A.H. (1990). Interaction of osteogenin, a heparin binding bone morphogenetic protein, with type IV collagen. *J Biol Chem* 265, 17281–84.

PARALKAR, V.M., VUKICEVIC, S., REDDI, A.H. (1991). Transforming growth factor β type 1 binds to collagen type IV of basement membrane matrix: Implications for development. *Dev Biol* 143, 303–9.

PARFITT, A.M. (1994). Osteonal and hemi-osteonal remodeling: The spatial and temporal framework for signal traffic in adult human bone. *J Cell Biochem* 55, 273–86.

PARFITT, A.M., MUNDY, G.R., ROODMAN, G.D., HUGHES, D.E., BOYCE, B.F. (1996). A new model for the regulation of bone resorption, with particular reference to the effects of bisphosphonates. *J Bone Miner Res* 11, 150–59.

REDDI, A.H. (1984). Extracellular matrix and development. In K.A. Piez and A.H. Reddi (eds.), *Extracellular Matrix Biochemistry.* New York: Elsevier.

REDDI, A.H. (1994). Symbiosis of biotechnology and biomaterials: Applications in tissue engineering of bone and cartilage. *J Cell Biochem* 56(2), 192–95.

REDDI A.H. (1997). Bone morphogenesis and modeling: Soluble signals sculpt osteosomes in the solid state. *Cell* 89, 159–61.

REDDI, A.H. (2000). Morphogenesis and tissue engineering of bone and cartilage: Inductive signals, stem cells, and biomimetic biomaterials. *Tissue Eng* 6(4), 351–59.

REDDI, A.H. (2005). BMPs: From bone morphogenetic to body morphogenetic proteins. *Cytokine Growth Factor Rev* 16(3), 249–50.

REDDI, A.H., HUGGINS, C. (1972). Biochemical sequences in the transformation of normal fibroblasts in adolescent rats. *Proc Natl Acad Sci USA* 69(6), 1601–5.

RIPAMONTI U. (2003). Osteogenic proteins of the transforming growth factor-$\beta$ superfamily. In H.L. Henry and A.W. Norman (eds.), *Encyclopedia of Hormones.* Academic Press, pp. 80–86.

RIPAMONTI, U. (2004). Soluble, insoluble and geometric signals sculpt the architecture of mineralized tissues. *J Cell Mol Med* 8(2), 169–80.

RIPAMONTI, U. (2005). Bone induction by recombinant human osteogenic protein-1 (hOP-1, BMP-7) in the primate *Papio ursinus* with expression of mRNA of gene products of the TGF-$\beta$ superfamily. *J Cell Mol Med* 9, 911–28.

RIPAMONTI, U. (2006). Soluble osteogenic molecular signals and the induction of bone formation. *Biomaterials* 27, 807–22.

RIPAMONTI, U. (2007). Recapitulating development: A template for periodontal tissue engineering. *Tissue Eng* 13, 51–71.

RIPAMONTI, U. (2009). Biomimetism, biomimetic matrices and the induction of bone formation. *J Cell Mol Med* 13(9B), 2953–72.

RIPAMONTI, U., CROOKS, J., KHOALI, L., RODEN L. (2009). The induction of bone formation by coral derived calcium carbonate/hydroxyapatite constructs. *Biomaterials* 30, 1428–39.

RIPAMONTI, U., CROOKS, J., MATSABA, T., TASKER, J. (2000a). Induction of endochondral bone formation by recombinant human transforming growth factor-$\beta_2$ in the baboon (*Papio ursinus*). *Growth Factors* 17(4), 269–85.

RIPAMONTI, U., DUARTE, R., FERRETTI, C. (2014). Re-evaluating the induction of bone formation in primates. *Biomaterials* 35(35), 9407–22.

RIPAMONTI, U., DUNEAS, N., VAN DEN HEEVER, B., BOSCH, C., CROOKS, J. (1997). Recombinant transforming growth factor-$\beta_1$ induces endochondral bone in the baboon and synergizes with recombinant osteogenic protein-1 (bone morphogenetic protein-7) to initiate rapid bone formation. *J Bone Miner Res* 12, 1584–95.

RIPAMONTI, U., FERRETTI, C. (2003). Mandibular reconstruction using naturally-derived bone morphogenetic proteins: A clinical trial report. In T.S. Lindholm (ed.), *Advances in Skeletal Reconstruction Using Bone Morphogenetic Proteins*. Singapore: World Scientific, pp. 277–89.

RIPAMONTI, U., FERRETTI, C., HELIOTIS, M. (2006). Soluble and insoluble signals and the induction of bone formation: Molecular therapeutics recapitulating development. *J Anat* 209, 447–68.

RIPAMONTI, U., HELIOTIS, M., FERRETTI, C. (2007). Bone morphogenetic proteins and the induction of bone formation: From laboratory to patients. *Oral and Maxillofacial Surg Clin North Am* 19, 575–89.

RIPAMONTI, U., HERBST, N.-N., RAMOSHEBI, L.N. (2005). Bone morphogenetic proteins in craniofacial and periodontal tissue engineering: Experimental studies in the non-human primate *Papio ursinus. Cytokine Growth Factor Rev* 16, 357–68.

RIPAMONTI, U., KLAR, R.M., RENTON, L.F., FERRETTI, C. (2010). Synergistic induction of bone formation by hOP-1 and TGF-$\beta$3 in macroporous coral-derived hydroxyapatite constructs. *Biomaterials* 31(25), 6400–10.

RIPAMONTI, U., MA, S., CUNNINGHAM, N., YATES, L., REDDI, A.H. (1992). Initiation of bone regeneration in adult baboons by osteogenin, a bone morphogenetic protein. *Matrix* 12, 202–12.

RIPAMONTI, U., RAMOSHEBI, L.N., MATSABA, T., TASKER, J., CROOKS, J., TEARE, J. (2001). Bone induction by BMPS/OPS and related family members in primates. *J Bone Joint Surg Am* 83A(Suppl 1, Pt 2), S116–27.

RIPAMONTI, U., RAMOSHEBI L.N., PATTON J., MATSABA T., TEARE J., RENTON L. (2004). Soluble signals and insoluble substrata: Novel molecular cues instructing the induction of bone. In E.J. Massaro and J.M. Rogers (eds.), *The Skeleton*. Humana Press, Totowa, New Jersey, pp. 217–27.

RIPAMONTI, U., RAMOSHEBI, L.N., TEARE, J., RENTON, L., FERRETTI, C. (2008). The induction of endochondral bone formation by transforming growth factor-$\beta_3$: Experimental studies in the non-human primate *Papio ursinus. J Cell Mol Med* 12(3), 1029–48.

RIPAMONTI, U., REDDI, A.H. (1995). Bone morphogenetic proteins: Applications in plastic and reconstructive surgery. *Adv Plastic Reconstr Surg* 11, 47–65.

RIPAMONTI, U., RODEN, L. (2010). The induction of bone formation by the transforming growth factor-$\beta$2 in the non-human primate *Papio ursinus* and its modulation by skeletal muscle responding stem cells. *Cell Prolif* 43, 207–18.

RIPAMONTI, U., TEARE, J., FERRETTI, C. (2012). A macroscopic bioreactor super activated by the recombinant human transforming growth factor-$\beta_3$. *Frontiers Physiol* 3, 172. doi: 10.3389/fphys.2012.00172, www.frontiersin.org.

RIPAMONTI, U., VAN DEN HEEVER, B., CROOKS, J., TUCKER, M.M., SAMPATH, T.K., RUGER, D.C. (2000b). Long term evaluation of bone formation by osteogenic protein-1 in the baboon and relative efficacy of bone-derived bone morphogenetic proteins delivered by irradiated xenogeneic collagenous matrices. *J Bone Miner Res* 15, 1798–809.

RIPAMONTI, U., VAN DEN HEEVER, B., HELIOTIS, M., DAL MAS, I., HAHNLE, U., BISCARDI, A. (2002). Local delivery of bone morphogenetic proteins using a reconstituted basenent membrane gel: Tissue engineering by Matrigel. *S Afr J Sci* 98, 429–33.

RIPAMONTI, U., VAN DEN HEEVER, B., SAMPATH, T.K., TUCKER, M.M., RUEGER, D.C., REDDI, A.H. (1996). Complete regeneration of bone in the baboon by recombinant human osteogenic protein-1 (bone morphogenetic protein-7). *Growth Factors* 13, 273–89.

ROBERTS, A.B., SPORN, M.B., ASSOIAN, R.K., SMITH, J.M., ROCHE, N.S., WAKEFIELD, L.M., HEINE, I.L., LIOTTA, L.A., FALANGA, V., KEHRL, J.H., FAUCI, A.S. (1986). Transforming growth beta: Rapid induction of fibrosis and angiogenesis in vivo and stimulation of collagen formation in vivo. *Proc Natl Acad Sci USA* 83(12), 4167–71.

SACERDOTTI, C., FRATTIN, G. (1901). Sulla produzione eteroplastica dell' osso. *Riv Accad Med Torino* 27, 825–36.

SAMPATH, T.K., COUGHLIN, J.E., WHETSTONE, R.M., BANACH, D., CORBETT, C., RIDGE, R.J., ÖZKAYNAK, E., OPPERMANN, H., RUEGER, D.C. (1990). Bovine osteogenic protein is composed of dimers of OP-1 and BMP-2A, two members of the transforming growth factor-beta superfamily. *J Biol Chem* 265(22), 13198–205.

SAMPATH, T.K., MUTHUKUMARAN, N., REDDI, A.H. (1987). Isolation of osteogenin, an extracellular matrix-associated, bone-inductive protein, by heparin affinity chromatography. *Proc Natl Acad Sci USA* 84(20), 7109–13.

SAMPATH, T.K., RASHKA, K.E., DOCTOR, J.S., TUCKER, R.F., HOFFMANN, F.M. (1993). Drosophila TGF-β superfamily proteins induce endochondral bone formation in mammals. *Proc Natl Acad Sci USA* 90, 6004–8.

SAMPATH, T.K., REDDI, A.H. (1981). Dissociative extraction and reconstitution of extracellular matrix components involved in local bone differentiation. *Proc Natl Acad Sci USA* 78(12), 7599–603.

SAMPATH, T.K., REDDI, A.H. (1983). Homology of bone-inductive proteins from human, monkey, bovine, and rat extracellular matrix. *Proc Natl Acad Sci USA* 80(21), 6591–95.

SCHNITZLER, C.M., RIPAMONTI, U., MESQUITA, J.M. (1993). Histomorphometry of iliac crest trabecular bone in adult male baboons in captivity. *Calcif Tissue Int* 52, 447–54.

SHAH, M., FOREMAN, D.M., FERGUSON, M.W. (1995). Neutralisation of TGF-beta 1 and TGF-beta 2 or exogenous addition of TGF-beta 3 to cutaneous rat wounds reduces scarring. *J Cell Sci* 108(Pt 3), 985–1002.

SHINOZAKI, M., KAWARA, S., HAYASHI, N., KAKINUMA, T., IGARASHI, A., TAKEHARA, K. (1997). Induction of subcutaneous tissue fibrosis in newborn mice by transforming growth factor beta—simultaneous application with basic fibroblast growth factor causes persistent fibrosis. *Biochem Biophys Res Commun* 240(2), 292–97.

SPAGNOLI, A., O'REAR, L., CHANDLER, R.L., GRANERO-MOLTO, F., MORTLOCK, D.P., GORSKA, A.E., WEIS, J.A., LONGOBARDI, L., CHYTIL, A., SHIMER, K., MOSES, H.L. (2007). TGF-beta signaling is essential for joint morphogenesis. *J Cell Biol* 177(6), 1105–17.

SPEMANN, H. (1938). *Embryonic Development and Induction.* New Haven, CT: Yale University Press.

TEN DIJKE, P., IWATA, K.K., GODDARD, C., PIELER, C., CANALIS, E., MCCARTHY, T.L., CENTRELLA, M. (1990). Recombinant transforming growth factor-beta 3: Biological activities and receptor-binding properties in isolated bone cells. *Mol Cell Biol* 10(9), 4473–79.

TRUETA, J. (1963). The role of the vessels in osteogenesis. *J Bone Joint Surg* 45B, 402–18.

TURING, A.M. (1952). The chemical basis of morphogenesis. *Philos R Soc Lond* 237, 37.

URIST, M.R. (1965). Bone: Formation by autoinduction. *Science* 150(698), 893–99.

URIST, M.R. (1968). The reality of a nebulous enigmatic myth. *Clin Orthop Rel Res* 59, 3–5.

URIST, M.R., DOWELL, T.A., HAY, P.H., STRATES, B.S. (1968). Inductive substrates for bone formation. *Clin Orthop Relat Res* 59, 59–96.

URIST, M.R., MCLEAN, F.C. (1952). Ostyeogenetic potency and new bone formation by induction in transplants to the anterior chamber of the eye. *J Bone Joint Surg Am* 34A, 443–76.

URIST, M.R., SILVERMAN, B.F., BURING, K., DUBUC, F.L., ROSENBERG, J.M. (1967). The bone induction principle. *Clin Orthop Relat Res* 53, 243–83.

VIOLA, J., LAL, B., GRAD, O. (2003). NSF: ABT report on the emergence of tissue engineering as a research field. http://www./nsf.gov/pub/2005/msf0450/start.htm.

VLODAVSKY, I., FOLKMAN, J., SULLIVAN, R., FRIDMAN, R., ISHAI-MICHAELI, R., SASSE, J., KLAGSBRUN, M. (1987). Endothelial cell-derived basic fibroblast growth factor: Synthesis and deposition into subendothelial extracellular matrix. *Proc Natl Acad Sci USA* 84, 2292–96.

VON HALLER, A. (1763). Experimentorum de ossium formatione. *In Opera Minora.* Francisci Grasset, Lausanne, p. 400.

VUKICEVIC, S., LUYTEN, F.P., KLEINMAN, H.K., REDDI, A.H. (1990). Differentiation of canalicular cell processes in bone cells by basement membrane matrix components: Regulation by discrete domains of laminin. *Cell* 63, 437–45.

WOZNEY, J.M., ROSEN, V., CELESTE, A.J., MITSOCK, L.M., WHITTERS, M.J., KRIZ, R.W., HEWICK, R.M., WANG, E.A. (1988). Novel regulators of bone formation: Molecular clones and activities. *Science* 242, 1528–34.

ZHENG, B., CAO, B., CRISAN, M., SUN, B., LI, G., LOGAR, A., YAP, S., POLLETT, J.B., DROWLEY, L., CASSINO, T., GHARAIBEH, B., DEASY, B.M., HUARD, J., PEAULT, B. (2007). Prospective identification of myogenic endothelial cells in human skeletal muscle. *Nat Biotechnol* 25, 1025–34.

# CHAPTER THREE

# Rapid Induction of Bone Formation by the Transforming Growth Factor-$\beta_3$ Isoform

*Ugo Ripamonti*

Bone Research Laboratory, School of Oral Health Sciences, Faculty of Health Sciences, University of the Witwatersrand, Johannesburg, Parktown, South Africa

## 3.1 The Transforming Growth Factor-$\beta_3$ Isoform

The biological significance of apparent redundancy is a very promising area of fertile research (Ripamonti et al. 2001, 2004, 2014a; Ripamonti 2004). The bone morphogenetic proteins (BMPs), together with the three mammalian and the amphibian transforming growth factor-$\beta$ isoforms (TGF-$\beta_1$, -$\beta_2$, -$\beta_3$, and -$\beta_5$, respectively), form a triad of subfamilies consisting of 40 or more homo- or heterodimeric structurally related morphogens that belong to the TGF-$\beta$ supergene family (Wozney et al. 1988; Kingsley 1994; Massagué 2000). All the listed morphogens of the family play pleiotropic roles in axial patterning, tissue morphogenesis, and organogenesis in both vertebrates and invertebrates (Wozney et al. 1988; Reddi 2000; Ripamonti et al. 2005; Ripamonti 2006).

The relationship between BMPs and TGF-$\beta$s is indicative of the biological significance of redundancy (Ripamonti et al. 1997) during mammalian embryogenesis and postnatal tissue induction and morphogenesis (Özkaynak et al. 1991; Luo et al. 1995; Vainio et al. 1993; Heikinheimo 1994; Helder et al. 1995; Thomadakis et al. 1999; Sampath et al. 1993; Piek et al. 1999; Ripamonti et al. 2006; Ripamonti 2007).

The initiation of bone formation involves a complex cascade of molecular and cellular processes that ultimately set the induction of bone in postnatal life as a recapitulation of embryonic development, and precisely lead to the development of multicellular highly organized structures with mineralized bone, osteoid (i.e., the newly synthesized matrix as yet to be mineralized), bone marrow formation, and hematopoiesis, all developmentally reengineered in the newly formed ossicles permeated by an array of several different cellular populations, including osteoblasts, osteocytes, osteoclasts, endothelial cells, and perivascular pericytes, together with stem cell niches in perivascular location (Reddi 2000; Ripamonti 2005, 2006, 2009; Zheng et al. 2007; Kovacic and Boehm 2008; Chen et al. 2009; Crisan et al. 2008; Caplan 2008; Benjamin et al. 1998).

We have highlighted that BMPs, pleiotropic members of the TGF-β supergene family, are an elegant example of Nature's parsimony in programming multiple specialized functions or pleiotropism, including the heterotopic induction of bone formation, deploying molecular isoforms with minor variation in amino acid motifs within highly conserved carboxy-terminal regions (Ripamonti 2003, 2004, 2006).

The unique prerogative of the induction of bone formation, originally solely ascribed to the bone morphogenetic proteins (Ripamonti et al. 1992) has been extended to additional TGF-β superfamily members (Ripamonti 2004, 2006; Ripamonti et al. 2004). These include decapentaplegic (DPP), 60A in *Drosophila melanogaster* (Sampath et al. 1993), and the three mammalian transforming growth factor-β (TGF-β) isoforms (Ripamonti et al. 1997, 2000, 2004, 2008, 2010; Ripamonti 2006; Ripamonti and Roden 2010a), the latter in primates only (Ripamonti et al. 2008, 2014a). The three mammalian TGF-β isoforms *per se* have shown a marked species, site, and tissue specificity of induction culminating in the rapid and substantial induction of bone formation in heterotopic intramuscular *rectus abdominis* sites. The induction of bone formation by the hTGF-$\beta_3$ isoform has also indicated that the third mammalian TGF-β isoform is the most powerful inductive morphogen so far tested in primates (Figures 3.1 and 3.2) (Ripamonti et al. 2008, 2010, 2014a; Klar et al. 2014).

The mammalian TGF-$\beta_3$ isoform is a member of the TGF-β supergene family of functionally diverse, but molecularly and structurally conserved, morphogens that regulate axial patterning, cell proliferation, differentiation and tissue induction, chemotaxis, and chemochinesis in a cell and organ context-specific manner (Ripamonti et al. 1997, 2008, 2010; ten Dijke et al. 1990; Ripamonti 2006).

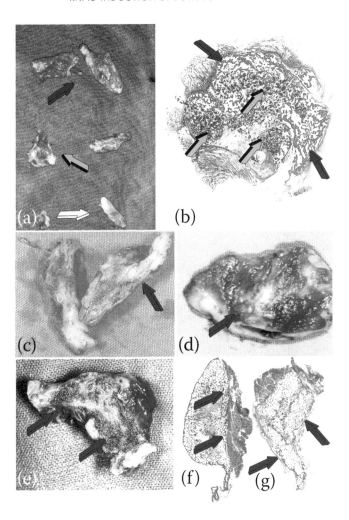

**FIGURE 3.1** Apparent redundancy of soluble molecular signals endowed with the striking prerogative to initiate the heterotopic induction of bone formation in the non-human primate Chacma baboon (*Papio ursinus*) 30 days after heterotopic implantation. (a) Intramuscular construction and fabrication of large heterotopic ossicles in the *rectus abdominis* muscle of adult non-human primates *P. ursinus* upon the implantation of 5 (white arrow), 25 (light blue arrow), and 125 (dark blue arrow) µg of recombinant human transforming growth factor-$\beta_3$ (hTGF-$\beta_3$) combined with allogeneic insoluble collagenous bone matrix as carrier. (b) Morphology of tissue induction in the *rectus abdominis* muscle by 5 µg of hTGF-$\beta_3$ with corticalization of the newly formed ossicle (dark blue arrows) surrounding scattered remnants of the insoluble bone matrix as carrier (light blue arrows). (c–e) Induction of large heterotopic corticalized ossicles (dark blue arrows) upon the implantation of 125 µg of hTGF-$\beta_3$ with (f, g) mineralized newly formed bone (dark blue arrows) surrounding osteoid matrix and scattered remnants of the collagenous matrix as carrier. Undecalcified sections embedded in K-Plast resin cut at 6 µm stained free-floating with Goldner's trichrome.

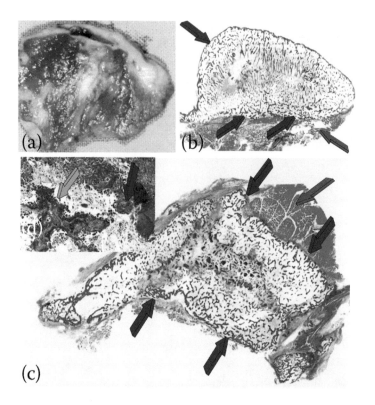

**FIGURE 3.2** Microphotographic images of the induction of bone formation by 125 µg of recombinant human transforming growth factor-$\beta_3$ (hTGF-$\beta_3$) recombined with allogeneic insoluble collagenous bone matrix 30 days after implantation in the *rectus abdominis* muscle. Corticalized newly formed mineralized bone (dark blue arrows) had generated within the surrounding muscle (magenta arrows in b and c). (d) Detail of hypercellular activity with large seams of a newly deposited osteoid, as yet to be mineralized, matrix (light blue arrow) populated by plumped contiguous osteoblasts actively secreting osteoid surfacing mineralized newly formed bone in blue (dark blue arrow). Undecalcified sections embedded in K-Plast resin cut at 5 µm stained free-floating with Goldner's trichrome.

The aim of this chapter is to convey a concise perspective on the induction of bone formation by the hTGF-$\beta_3$ isoform when implanted in the *rectus abdominis* muscle of *Papio ursinus* with either the insoluble collagenous bone matrix carrier or coral-derived macroporous bioreactors, the latter used to identify some of the mechanisms of the biological activity of the hTGF-$\beta_3$ in primates' heterotopic sites (Ripamonti et al. 2014a, 2015; Klar et al. 2014).

Our latest investigations on the substantial induction of bone formation by the hTGF-$\beta_3$ isoform have raised the need to reevaluate our present understanding of the induction of bone formation in primate species (Ripamonti et al. 2014a, 2015), additionally highlighting the central biological mechanisms that frame the fundamental mechanisms of human biology (Uniquely Human Biology 2014).

## 3.2 TGF-$\beta_3$ Isoform and the Induction of Bone Formation in Primates

Implantation of doses of hTGF-$\beta_3$ combined with allogeneic insoluble collagenous bone matrix in the *rectus abdominis* muscle of *P. ursinus* results in the induction of large corticalized ossicles by day 30 (Figures 3.1 and 3.2). Doses of 5, 25, and 125 µg of hTGF-$\beta_3$ initiate the construction of large corticalized ossicles embedded within the *rectus abdominis* muscle (Figures 3.1 and 3.2). Generated ossicles show a planar convex geometry extending for several centimeters along the longitudinal plane of the fascia. Cut surfaces show peripheral corticalization, while scattered remnants of the collagenous matrix occupy the center of the specimens, together with a highly vascular invading connective tissue matrix (Figure 3.2). Morphological analyses show continuous osteoblastic cell differentiation, osteoid synthesis together with prominent angiogenesis, and capillary sprouting (Figure 3.2d).

Interestingly, however, equal or superior doses of the recombinant morphogen when implanted in orthotopic calvarial sites of the same animals and harvested at the same time period (Figure 3.3) show minimal osteogenesis confined to the margins of the defects only (Figure 3.3a and b) (Ripamonti et al. 2008, 2009a). Newly formed bone, albeit limited, is strictly confined to the margins of the craniotomies only (Figure 3.3b, d, and e). Two specimens treated with 125 µg of hTGF-$\beta_3$ showed islands of newly formed mineralized bone across the implanted collagenous matrix as carrier (Figure 3.3f) (Ripamonti et al. 2008) and below the temporalis muscle. On day 90, bone formation in calvarial defects treated with the hTGF-$\beta_3$ osteogenic device remained limited, with scattered areas of osteogenesis below the pericranium and the temporalis muscle. Isolated specimens showed the induction of bone pericranially across the treated defects (Figure 3.4a) (Ripamonti et al. 2008, 2009a), replicating an identical pattern of tissue induction and morphogenesis as shown by calvarial specimens treated with 100 µg of recombinant hTGF-$\beta_2$ (Figure 3.4b) (Ripamonti et al. 2000; Ripamonti and Roden 2010a).

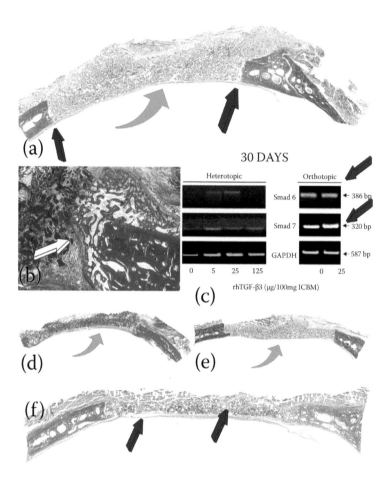

**FIGURE 3.3** Site/tissue specificity of the induction of bone formation by the recombinant human transforming growth factor-$\beta_3$ (hTGF-$\beta_3$) recombined with 1 g of allogeneic insoluble collagenous bone matrix implanted in nonhealing calvarial defects of *Papio ursinus* and harvested on day 30 after orthotopic implantation. (a) Lack of bone induction and differentiation upon calvarial implantation of the recombinant morphogen (light blue arrow). (b) Limited induction of bone formation at the craniotomy edge only (white arrow); newly formed bone blends into the remnants of the collagenous matrix as carrier, but is separated from the surrounding tissues by a prominent fibrous layer (white arrow) inhibiting the induction of bone formation. Morphology in (d) directly correlates to PCR analyses (c) showing on day 30 overexpression of *Smad-6* and *-7* in orthotopic sites (dark blue arrows), but not in heterotopic sites (left panel, c). (d, e) Reproducibility of complete lack of bone differentiation (light blue arrows) by the hTGF-$\beta_3$ osteogenic devices in additional specimens harvested and prepared from different implanted calvariae. (f) Calvarial defect harvested on day 30 after implantation of 125 µg of hTGF-$\beta_3$ recombined with thoroughly morcellated fragments of autogenous *rectus abdominis* muscle (Ripamonti et al. 2008, 2009a) (dark blue arrows). Undecalcified sections embedded in K-Plast resin cut at 5 µm, stained free-floating with Goldner's trichrome.

Morphological analyses on day 90 showed a recurrent pattern of histological features extending from the pericranial to the endocranial surfaces of the specimens, with particular reference to the interfacial region (Figure 3.4a and b). Of note, 125 μg of hTGF-$\beta_3$ osteogenic devices additionally treated with autogenous morcellated fragments of *rectus abdominis* muscle showed the induction of chondrogenesis between newly formed mineralized bone and the collagenous matrix as carrier (Figure 3.4c). The addition of morcellated fragments of autogenous *rectus abdominis* muscle segments engineered endochondral bone formation in calvarial membranous bone with large islands of chondrogenesis as a recapitulation of embryonic development 90 days after calvarial implantation of the hTGF-$\beta_3$ osteogenic device (Figure 3.4c) (Ripamonti et al. 2008, 2009a).

The presence of inhibitory binding proteins or the expression of inhibitory Smad proteins is highly suggested by the morphological pattern of tissue induction of calvarial specimens treated with the hTGF-$\beta_3$ osteogenic device (Figures 3.3 and 3.4). Bone that had formed at both interfacial regions seemed to be morphologically inhibited to further growth, with centripetal expansion resulting instead in limited bone formation across the pericranial aspect of the defects, with lack of bone formation endocranially along the dural surface (Figure 3.4a and b).

The remarkable induction of bone formation by the hTGF-$\beta_3$ osteogenic device in rectus abdominis heterotopic sites of *P. ursinus* has suggested that the limited induction of bone formation in orthotopic calvarial sites is the result of inhibitory morphogens or limited responding stem cells at the site of orthotopic calvarial implantation (Ripamonti et al. 2008, 2009a).

Both molecular and cellular hypotheses were tested by reverse transcription polymerase chain reaction (RT-PCR) analyses of hTGF-$\beta_3$-treated calvarial defects with or without the addition of morcellated fragments of *rectus abdominis* muscle. Thoroughly minced, morcellated fragments of autogenous *rectus abdominis* muscle were added to hTGF-$\beta_3$ osteogenic devices reconstituted with allogeneic insoluble collagenous bone matrices as carrier (Ripamonti et al. 2008, 2009a).

Molecularly, the observed limited induction of the hTGF-$\beta_3$ osteogenic device in calvarial defects of *P. ursinus* is due to the overexpression of *Smad-6* and *-7* downstream antagonists of the TGF-$\beta$ signaling pathway (Ripamonti et al. 2008, 2009a); RT-PCRs of tissue specimens generated by hTGF-$\beta_3$ demonstrated robust expression of *Smad-6* and *-7* in orthotopic calvarial sites on day 30, with limited expression, if any, in heterotopic *rectus abdominis* sites (Figure 3.3c) (Ripamonti et al. 2008, 2009a).

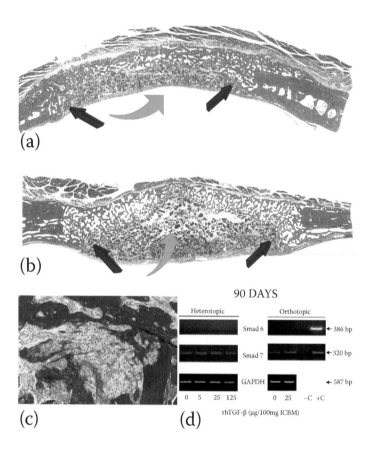

**FIGURE 3.4** Morphology of calvarial regeneration and induction of bone formation in calvaria defects implanted with (a) 125 μg of human recombinant transforming growth factor-$\beta_3$ (hTGF-$\beta_3$) and (b) 100 μg of hTGF-$\beta_2$ (Ripamonti et al. 2000) recombined with 1 g of allogeneic insoluble collagenous bone matrix as carrier and harvested on day 90 after implantation. There is induction of bone formation, albeit pericranially only, using both mammalian isoforms on day 90. (a, b) The interfacial areas of both specimens show the limited induction from the recipient calvarial margins; arrows (dark blue arrows in a and b) point to the inhibition of bone formation from the severed calvarial edges endocranially. New bone formation is inhibited to growth centrifugally from the endo-cranial calvarial cuts (dark blue arrows); morphology as shown in (a) and (b) is directly correlated to the PCR analyses as shown in (d), illustrating limited expression of *Smad-6* and -7 in orthotopic sites on day 90, compared to day 30, therefore resulting in limited induction of bone formation under the pericranium by day 90 (a, b). (c) High-power view of the induction of chondrogenesis (light blue arrow) in calvarial membranous bone implanted with 125 μg of hTGF-$\beta_3$ in morcellated fragments of autogenous *rectus abdominis* muscle. The addition of morcellated fragments of autogenous *rectus abdominis* muscle engineers endochondral bone differentiation with large islands of chondrogene-sis (light blue arrow) as a recapitulation of embryonic development 90 days after implan-tation of the osteogenic device in nonhealing calvarial defects. The induction of bone

Previous studies have suggested that overexpression of *Smad-6* and *-7* in treated calvarial defects may be due to the vascular endothelial network of the arachnoids (Topper et al. 1997) expressing signaling proteins modulating the expression of the inhibitory Smads in pre-osteoblastic and osteoblastic calvarial cell lines, controlling the induction of bone formation in the primate calvarium (Figures 3.3 and 3.4) (Ripamonti et al. 2008, 2009a).

Of interest for tissue engineering in clinical contexts, partial restoration of the induction of bone formation by hTGF-$\beta_3$ has been obtained by combining the hTGF-$\beta_3$ osteogenic device with morcellated, thoroughly minced fragments of autogenous *rectus abdominis* muscle prior to the implantation in calvarial defects (Figure 3.5) (Ripamonti et al. 2008, 2009a).

The observation that morcellated fragments of autogenous *rectus abdominis* muscle partially restored the bone induction activity of the hTGF-$\beta_3$ isoform has been replicated to periodontal regenerative studies in which harvested autogenous fragments of rectus abdominis muscle were finely minced and added to 75 µg of hTGF-$\beta_3$ in a Matrigel matrix implanted in surgically created Class II and III furcation defects of *P. ursinus* (Ripamonti and Petit 2009; Ripamonti et al. 2009b). As discussed in Chapter 7, the direct application of 75 µg of hTGF-$\beta_3$ in a Matrigel matrix, together with finely morcellated autogenous fragments of rectus abdominis muscle, resulted in greater alveolar bone formation and induction of cementogenesis along surgically exposed root surfaces (Ripamonti and Petit 2009; Ripamonti et al. 2009b).

Of note, the elevated expression of *Smad-6* and *-7* in orthotopic calvarial specimens seen on day 30 (Figure 3.3c) was no longer observed on day 90 (Figure 3.4d). Importantly, supporting the mechanistic insights of the induction of bone formation in calvarial sites, the relative reduction of expression of *Smad-6* and *-7* in calvarial sites as shown on day 90 (Figures 3.4d) correlated with the induction of bone formation, albeit limited pericranially on day 90 (Figure 3.4a and b).

---

**FIGURE 3.4** *Continued*

formation by both mammalian morphogens strongly indicates the presence of inhibitory proteins that inhibit the induction of bone formation from the severed calvariae with centripetal extension. This results in lack of bone formation endocranially above the dura mater and limited bone formation across the pericranial aspect of the defects. The pattern of tissue morphology is highly suggestive of a radiating inhibitory activity from the dural layer or from the highly vascular arachnoid spaces. Indeed, studies have shown MAD-related (mothers against decapentaplegic) genes that are selectively inducible by vascular endothelium below the dura (Topper et al. 1997).

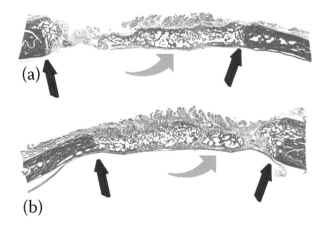

**FIGURE 3.5** Inductive effect of autogenous morcellated, finely minced *rectus abdominis* muscle fragments. The implantation of morcellated, finely minced, striated autogenous muscle fragments, together with 125 μg of recombinant human transforming growth factor-β₃ (hTGF-β₃) implanted in calvarial defects (dark blue arrows) of *Papio ursinus* harvested on day 90 after implantation restores the induction of bone formation of the implanted hTGF-β₃ isoform. (a, b) Morphology of calvarial regeneration with partial restoration of the inductive activity of the recombinant morphogen by the myoblastic or pericytic perivascular stem cells with the morcellated preparations of autogenous *rectus abdominis* muscle. Substantial induction of newly formed mineralized bone (large light blue arrows) across the defects. Undecalcified sections embedded in K-Plast resin cut at 6 μm, stained free-floating with Goldner's trichrome.

### 3.3 Coral-Derived Macroporous Bioreactors as Delivery System for the Biological Activity of the Recombinant hTGF-β₃ Isoform

The Bone Research Laboratory has shown that in primates, and in primates only, doses of the recombinant human transforming growth factor-β₃ (hTGF-β₃), when combined with an allogeneic insoluble collagenous bone matrix as carrier, result in the induction of large corticalized ossicles by days 30 and 90 after implantation in the *rectus abdominis* muscle of *P. ursinus* (Figures 3.1 and 3.2). Combining 125 μg doses of hTGF-β₃ to coral-derived macroporous constructs (Ripamonti et al. 2012a, 2014a, 2014b; Klar et al. 2014) results in substantial induction of bone formation (Figure 3.6). On days 60 and 90, the induction of bone formation forms across the macroporous spaces tightly attached to the implanted substratum (Figure 3.6a), with

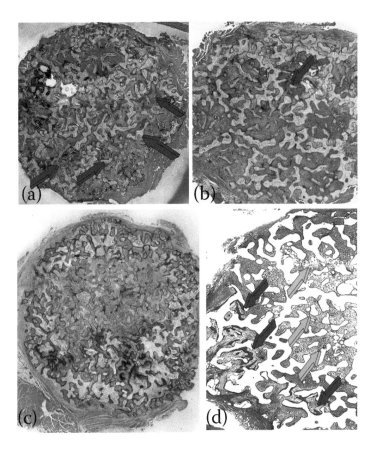

**FIGURE 3.6** The induction of bone formation by the recombinant human transforming growth factor-$\beta_3$ (hTGF-$\beta_3$) when recombined with coral-derived macroporous bioreactors implanted in heterotopic sites of the *rectus abdominis* muscle of the Chacma baboon (*Papio ursinus*) (Ripamonti et al. 2012, 2014a, 2014b; Klar et al. 2014). (a) Induction of bone formation (dark blue arrows) by 125 µg of hTGF-$\beta_3$ within the coral-derived macroporous spaces. (b) Marked inhibition of the induction of bone formation with only one island of bone (dark blue arrow) in coral-derived macroporous bioreactors preloaded with binary application of 125 µg of hTGF-$\beta_3$ and 150 µg of recombinant human Noggin (hNoggin). (c) Complete inhibition of bone formation in another specimen treated with binary application of 125 µg of hTGF-$\beta_3$ and 150 µg of hNoggin with inactive fibrovascular tissue invading the macroporous spaces (Ripamonti et al. 2014b). (d) Limited bone formation at the periphery of the macroporous construct only in a specimen pretreated with binary application of 125 µg of hTGF-$\beta_3$ and 125 µg of hNoggin, respectively. Minor bone formation by induction at the periphery only with islands of poorly inductive mesenchymal fibrovascular tissue within the macroporous spaces of the implanted bioreactor. (a–c) Undecalcified sections cut with the diamond cutting and grinding Exakt saw at 27 µm.

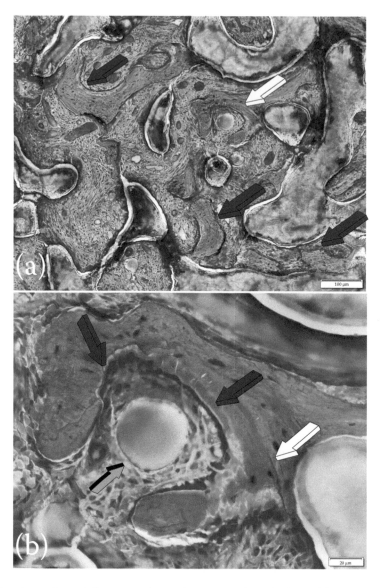

**FIGURE 3.7** Tissue induction and morphogenesis of newly formed bone by 125 μg of recombinant human transforming growth factor-β$_3$ (hTGF-β$_3$) reconstituted with coral-derived macroporous bioreactors and implanted in the *rectus abdominis* muscle of the Chacma baboon (*Papio ursinus*). Florid induction of bone morphogenesis extending within the macroporous spaces; newly formed trabeculae are covered by osteoblastic cells secreting newly formed bone matrix (Ripamonti et al. 2014b). (b) High-power view highlighting the induction of bone formation by the osteogenetic and morphogenetic vessels of Aristotle's and Trueta's definitions (Crivellato et al. 2007; Trueta 1963), which initiate the plastic morphogenesis of bone around the morphogenetic vessel (light blue arrow) with the bone surfaced by multiple osteoblastic-like cells (dark blue arrows)

the induction of trabeculations across the treated macroporous spaces (Figure 3.7).

Mechanistically, the induction of bone formation by the hTGF-$\beta_3$ isoform when preloaded onto coral-derived macroporous bioreactors is abolished by preloading the macroporous constructs with 125 (Klar et al. 2014) or 150 (Ripamonti et al. 2014a, 2014b) μg of recombinant hNoggin (Figure 3.6b–d). The lack of bone formation by induction in macroporous coral-derived bioreactors pretreated with binary applications mechanistically indicates that the induction of bone formation as initiated by the hTGF-$\beta_3$ isoform is set into motion by *BMP* gene expression upon heterotopic implantation of hTGF-$\beta_3$, and thus blocked by hNoggin, a BMP antagonist (Groppe et al. 2003; Gazzerro et al. 1998).

Increasing doses of the hTGF-$\beta_3$ up to 250 μg results in prominent and significant induction of bone formation *de facto* entirely localized outside the profile of the implanted pretreated macroporous constructs, resulting in the superactivation of the heterotopically implanted bioreactors (Figure 3.8). Superactivated bioreactors show the prominent induction of bone formation 2–3 cm away from the periphery of the implanted macroporous constructs (Figure 3.8e) (Ripamonti et al. 2012a).

The latter findings indicate the establishment of a rapid TGF-$\beta_3$ morphogen gradient (Lander 2007) so far unreported as yet by any gene product of the TGF-$\beta$ superfamily, possibly further potentiated by cell-to-cell receptor activation (Lander 2007) and by the establishment of a TGF-$\beta_3$ morphogen gradient by long-range activation (Michael Jones and Smith 1998).

The remarkable and substantial induction of bone formation well outside the profile of the heterotopically implanted bioreactors with limited, if any, bone formation by induction within the central and internal macroporous spaces (Figure 3.8a and e) has raised the suggestion that such an extended range of biological activity is due to the establishment of a diffusion gradient (Lander 2007) or to the induction of a sequential chain of cellular induction that results in the rapid initiation of bone formation 2–3 cm away from the implanted macroporous bioreactors (Figure 3.9e) (Ripamonti et al. 1997, 2012a), rapidly recruiting intra–*rectus abdominis* muscle stem

---

**FIGURE 3.7** *Continued*

surrounding the morphogenetic vessels together with tensional fibers (white arrow in b), orchestrating the plasticity of the pattern of tissue morphogenesis around the central osteogenetic/morphogenetic vessel. Undecalcified sections cut with the diamond cutting and grinding Exakt saw at 27 μm.

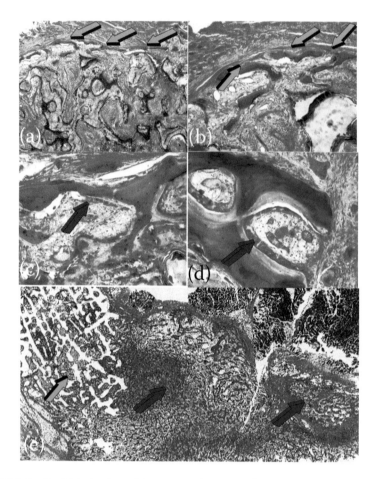

**FIGURE 3.8** Rapid induction of tissue morphogenesis by coral-derived macroporous bioreactors preloaded with 250 µg of recombinant human transforming growth factor-$\beta_3$ (hTGF-$\beta_3$) harvested on day 60 after intramuscular implantation. (a, b) The induction of bone formation is initiated only at the very periphery of the implanted bioreactors (light blue arrows), with lack of bone induction within the internal/central areas of the preloaded coral-derived bioreactors. (c, d) Newly formed bone at the periphery of the macroporous bioreactor showing mineralized bone covered by osteoid seams populated by contiguous osteoblasts (dark blue arrows) in close proximity to invading sprouting capillaries. (e) Prominent and significant induction of bone formation *de facto* entirely localized outside the profile of the implanted coral-derived macroporous construct pre-treated with 250 µg of hTGF-$\beta_3$ and harvested 20 days after intramuscular implantation. Bone forms by induction within the *rectus abdominis* muscle at considerable distance (dark blue arrows) from the implanted preloaded bioreactor, which shows lack of bone differentiation within the internal macroporous spaces (light blue arrow) (Ripamonti et al. 2012a, 2014a). Undecalcified sections cut with the diamond cutting and grinding Exakt saw at 27 µm.

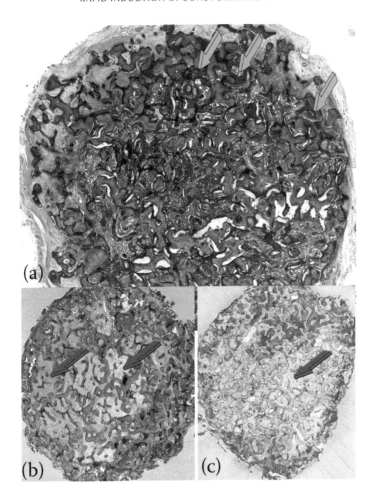

**FIGURE 3.9** The spontaneous induction of bone formation by coral-derived macroporous constructs (Ripamonti et al. 1993) is abolished after preloading coral-derived macroporous constructs with 125 μg of hNoggin (Klar et al. 2013; Ripamonti et al. 2014a, 2014b). (a) Induction of bone formation (light blue arrows) by the coral-derived macroporous bioreactor implanted solo within the *rectus abdominis* muscle of *Papio ursinus*. (b, c) Complete absence of bone induction by coral-derived macroporous constructs preloaded with 125 μg of hNoggin; lack of bone differentiation throughout the macroporous spaces (magenta arrows). Undecalcified sections cut with the diamond cutting and grinding Exakt saw at 27 μm.

cells surrounding the macroporous constructs directly transformed into secreting osteoblasts (Ripamonti et al. 2014a) long before responding stem cell invasion within the peripheral macroporous spaces. The morphological observations once again support the statement that "all organizations in mammalian tissues start at the periphery" (Martin Hale November

2014, personal communication). Tissue transformation at the periphery of the implanted macroporous constructs may be the result of an accelerated hTGF-$\beta_3$ specific diffusion gradient that rapidly transforms all available stem and progenitor cells at the periphery of the implanted superactivated hTGF-$\beta_3$ osteogenic macroporous bioreactors.

Of note, coral-derived macroporous bioreactors are endowed with the striking capacity to induce the morphogenesis of bone when implanted in heterotopic sites of the *rectus abdominis* muscle of *P. ursinus*, even without the exogenous application of the osteogenic proteins of the TGF-$\beta$ supergene family (Figure 3.9a) (Ripamonti 1991, 2000, 2006, 2009; Ripamonti et al. 1993, 2009c; Ripamonti and Roden 2010b; Klar et al. 2013).

Importantly, as it reveals the mechanistic insights of the induction of bone formation by the hTGF-$\beta_3$ isoform (Ripamonti et al. 2014a, 2015; Klar et al. 2014), the addition of 125 or 150 µg of recombinant hNoggin to coral-derived macroporous bioreactors abolishes the induction of bone formation, as shown molecularly (Klar et al. 2013; Ripamonti et al. 2015) and morphologically (Figure 3.9b and c) (Ripamonti et al. 2014a).

The inhibition of the induction of bone formation by either 125 (Klar et al. 2013) or 150 (Ripamonti et al. 2014b) µg of hNoggin has indicated that the induction of bone formation as initiated by the coral-derived macroporous bioreactors is set into motion via the *BMP* pathway (Figure 3.9) (Ripamonti et al. 2010, 2014a; Klar et al. 2013). Importantly, binary application of 125 µg of hTGF-$\beta_3$ and 125 or 150 µg of hNoggin resulted in the inhibition of the induction of bone formation within the macroporous spaces (Figure 3.9b and c), as Noggin is a strong BMP antagonist (Groppe et al. 2003; Gazzerro et al. 1998).

The morphological and molecular analyses thus reveal that the intrinsic and spontaneous induction of bone formation by untreated coral-derived bioreactors (Figure 3.9a) (Ripamonti 1991; Ripamonti et al. 1993) with or without doses of the hTGF-$\beta_3$ isoform is *via* the BMP pathway (Figure 3.6a), and that the addition of hNoggin profoundly inhibits the molecular and biological cascades as set by the expressed *BMP* genes, resulting in a lack of bone differentiation by induction (Figures 3.9b and c and 3.10) (Klar et al. 2014; Ripamonti et al. 2014a, 2015) in both untreated and hTGF-$\beta_3$-treated bioreactors (Figures 3.6b–d and 3.9b and c).

The lack of bone differentiation within hTGF-$\beta_3$-pretreated macroporous spaces with substantial induction at the periphery of the implanted constructs replicates the induction of bone formation and the maturational gradient of the synergistic induction of bone formation (Ripamonti et al. 1997). Heterotopic implantation of 25 µg of recombinant human osteogenic protein-1 (hOP-1) in binary application with 0.5 or 1.5 µg of hTGF-$\beta_1$ results in rapid tissue induction and morphogenesis transforming the *rectus abdominis* muscle into bone (Figure 3.11) (Ripamonti et al. 1997, 2014a). The morphological analyses of newly formed ossicles generated by synergistic binary application have suggested constraints on ossicle growth as a result of limited central vascular invasion, together with lack of stem cell or responding cell invasion within the implanted carrier surrounded by the surgically severed intramuscular tissue (Ripamonti et al. 1997).

In spite of the morphogenetic constraints because of limited angiogenesis, stem cell migration, and capillary invasion into the heterotopically implanted osteogenic devices, binary synergistic applications of 20:1 hOP-1/hTGF-$\beta_1$ (Ripamonti et al. 1997) resulted in the *rectus abdominis* muscle's transfiguration *in vivo* by day 15 with trabeculae of newly formed mineralized bone covered by contiguous secreting plumped osteoblastic cells rapidly displacing and transforming muscle fibers into bone (Figure 3.11), together with the early initiation of bone marrow formation and hematopoiesis (Figure 3.11).

The lack of responding stem cell invasion is due to the rapid transformation with tissue induction and morphogenesis at the periphery of the implanted macroporous bioreactors recruiting most of the available responding cells, thus blocking further cellular migration within the implanted osteogenetic matrix. Precursor stem cells are activated by waves of the hTGF-$\beta_3$ diffusion gradients across the macroporous spaces and the surrounding adjacent muscle (Ripamonti et al. 2015). The rapid induction of bone formation at the very periphery of the macroporous constructs "consumes" all the available responding cells, resulting in the morphological presence of "voids" in the center of large ossicles when using insoluble collagenous bone matrix as carrier (Ripamonti et al. 1997). Importantly, voids with limited, if any, tissue induction and morphogenesis within the central and internal areas of the macroporous bioreactors also relate to the rapid tissue induction and transformation of responding stem cells at the periphery of the implanted hTGF-$\beta_3$-pretreated macroporous constructs (Figure 3.3) (Ripamonti et al. 2014a, 2015).

The third mammalian isoform combined with either insoluble collagenous bone matrix or coral-derived macroporous bioreactors morphologically equates the synergistic induction of bone formation whereby a single human recombinant osteogenic protein is significantly superactivated three- to fourfold by the

(a)

### Relative *RUNX-2* and *OC* expression in pre-treated coral derived constructs at 15 days

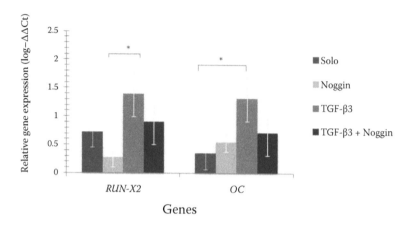

(b)

### Relative *BMP-2* and *TGF-β₃* expression in pre-treated coral derived constructs *in vivo*

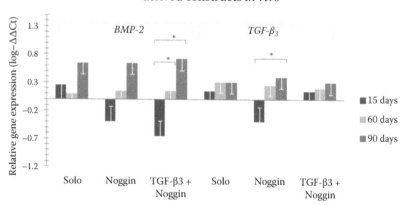

**FIGURE 3.10** Relative changes in gene expression of *RUN-X2, Osteocalcin, BMP-2,* and *TGF-β3* in untreated coral-derived 7% hydroxyapatite/calcium carbonate bioreactor (HA/CC) untreated controls, hNoggin, or hTGF-β₃-pretreated macroporous bioreactors (Ripamonti et al. 2014a). (a) *RUN-X2* and *Osteocalcin* expression is increased relative to control *solo* specimens at 15 days, and this correlates to the rapid induction of bone formation as seen between 15 and 30 days postimplantation in the *rectus abdominis* muscle (Ripamonti et al. 2014a, 2014b; Klar et al. 2014). The pronounced expression of *RUN-X2* and *Osteocalcin* has indicated recruitment and differentiation of mesenchymal invading stem cell progenitors, together with dedifferentiation of

addition of relatively low doses of a hTGF-β isoform (Figure 3.11) (Ripamonti et al. 1997, 2014a; Ripamonti 2004, 2006).

The rapid induction of bone formation by the hTGF-β₃ osteogenic device delivered by insoluble collagenous bone matrix or coral-derived macroporous bioreactors, together with *hTGF-β3, BMP-3, BMP-2, OP-1, collagen type IV, Osteocalcin, RUNX-2* expression, hypercellular osteoblastic activity, osteoid synthesis, angiogenesis, and capillary sprouting, is the novel molecular and morphological basis for the induction of bone formation in clinical contexts (Ripamonti et al. 2008, 2014a).

On day 15, harvested hTGF-β₃-pretreated coral-derived macroporous bioreactors showed the differentiation of fibrin/fibronectin rings expanding within the macroporous spaces structurally organizing tissue patterning and morphogenesis (Figure 3.12) (Ripamonti et al. 2014a; Klar et al. 2014). Advancing extracellular matrix rings provide the structural anchorage for hyperchromatic cells interpreted as differentiating osteoblasts reprogrammed by the hTGF-β₃ isoform from invading myoblastic/pericytic differentiated cells (Figure 3.12). *RUNX-2* and *Osteocalcin* expressions are significantly upregulated in hTGF-β₃-treated bioreactors on day 15 (Figure 3.10), supporting the morphological observation of invading cells differentiating into the osteoblastic phenotype with hypercellular osteoblastic activity and extracellular matrix secretion (Figure 3.12d).

The images presented in Figure 3.12 indicate the extent of structural organization of the extracellular matrix directly modulated by the hTGF-β₃ isoform (Ripamonti et al. 2014a; Klar et al. 2014). High-power digital images indicate that relatively high doses of 125 μg of hTGF-β₃ directly reprogram resident invading recruited myoblastic or pericytic/perivascular cells into highly active secreting osteoblasts depositing bone matrix by day 15, later to be mineralized by day 17, as shown by a time study in the non-human primate *P. ursinus* (Ripamonti 2014, Bone Research Laboratory, unpublished data). Importantly, for the subsequent induction of bone formation, the induction of advancing

---

**FIGURE 3.10** *Continued*

differentiated committed myoblastic or pericytic perivascular cells (Ripamonti et al. 2014a, 2014b). (b) Relative gene expression profile of *BMP-2* and *TGF*-β₃ in untreated 7% HA/CC controls, hNoggin, and TGF-β₃/hNoggin-treated devices at days 15, 60, and 90 after heterotopic *rectus abdominis* implantation. There is downregulation of *BMP-2* in hNoggin-treated macroporous bioreactors at day 15; of note, *BMP-2* is also downregulated in macroporous constructs pretreated with binary hNoggin/hTGF-β₃ applications, indicating the direct effect of 125 μg of hNoggin on *BMP* expression, correlating to the limited induction of bone formation within the macroporous spaces.

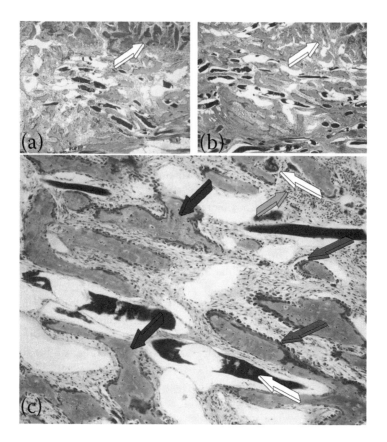

**FIGURE 3.11** Tissue transfiguration and the rapid induction of bone formation by the synergistic induction of bone formation (Ripamonti et al. 1997, 2010). Intra–*rectus abdominis* implantation of a 20:1 ratio of hOP-1/hTGF-$\beta_1$ by weight, 25 and 1.5 µg, respectively, results in the rapid induction of bone formation transfiguring the *rectus abdominis* muscle (white arrows in a and b) and its fibers into mineralized bone by day 15. (c) Bone mineralizes (dark blue arrows) and newly formed mineralized bone is covered by continuous layers of plumped highly active secreting osteoblasts actively depositing osteoid, as yet to be mineralized. (c) White arrows indicate fragmented transformed muscle fibers embedded in the highly active ossicle with hematopoietic bone marrow differentiation (light blue arrow in c) within the newly formed heterotopic ossicle by day. Undecalcified sections embedded in K-Plast resin cut at 6 µm stained free-floating with Goldner's trichrome.

extracellular matrix rings within the macroporous bioreactors superactivated by the hTGF-$\beta_3$ isoform provides structural anchorage for hyperchromatic osteoblastic-like cells reprogrammed by the hTGF-$\beta_3$ from invading myoblastic/pericytic cells (Figure 3.12) (Klar et al. 2014; Ripamonti et al. 2014a, 2014b), later differentiating in bone-forming trabeculations covered by osteoid seams populated by contiguous hyperchromatic

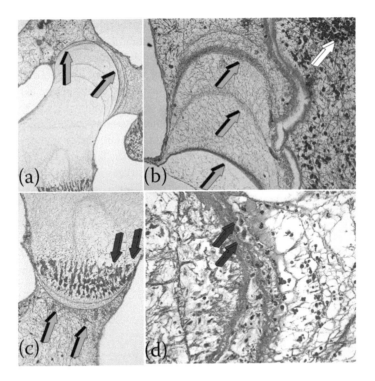

**FIGURE 3.12** Extracellular matrix induction and tissue patterning on day 15 after heterotopic intramuscular *rectus abdominis* implantation in *Papio ursinus* by super-activated coral-derived macroporous bioreactors pretreated with 125 µg of recombinant human transforming growth factor-$\beta_3$ (hTGF-$\beta_3$) (Ripamonti et al. 2014a; Klar et al. 2014). (a, b) Differentiation, alignment, and expansion of fibrin/fibronectin rings within the macroporous spaces of the activated bioreactors (light blue arrows in a and b). Differentiated fibrin/fibronectin rings expanding within the macroporous spaces structurally organize tissue patterning and morphogenesis, predating the induction of bone formation. The highly organized and expanding fibrin/fibronectin rings are endowed with tractional forces, as shown by patterned collagenic fibers across concavities constructing mesenchymal tissue bridges depicting tractional patterned forces within cells across the concavity borders (Ripamonti et al. 2012b). High-power digital images show that the expanding fibrin/fibronectin rings compress extracellular fluid and newly formed matrix with cellular elements against the coral-derived macroporous surfaces, at the same time transporting within the interlaced matrix differentiating cells (dark blue arrows in c and d) that initiate matrix synthesis and the early differentiation of bone (dark blue arrows in d), which rapidly mineralize by day 17 after heterotopic implantation (Ripamonti et al. 2014b).

osteoblasts by day 17 (Ripamonti 2014, Bone Research Laboratory, unpublished data).

Bone then forms rapidly between 15 and 30 days, initiating at the periphery of the implanted macroporous bioreactors, even at some distance from the coral-derived constructs. We have shown that the primary differentiating events that set the induction of bone formation within untreated coral-derived macroporous bioreactors primarily develop within the macroporous spaces after stem cell migration and differentiation onto nanotopographically modified macroporous surfaces by osteoclastogenesis, with lack of *BMP-2* or other transforming genes' expression in the surrounding adjacent muscle (Ripamonti et al. 2015).

The induction of bone must thus proceed via surface modifications of the highly crystalline calcium carbonate/hydroxyapatite bioreactor, which, as suggested in previous work (Ripamonti et al. 2014a), is the "self-constructor" of the induction of bone formation *via* invocation of $Ca^{++}$ release, osteoclastic topographical surface modifications, induced angiogenesis, and stem cell differentiation into osteoblastic-like cells expressing *BMP-2* and related genes, including *TGF-β* genes, later secreting biologically active gene products to be embedded into the macroporous, biologically active surfaces initiating the induction of bone formation as a secondary response (Figure 3.10a) (Klar et al. 2013; Ripamonti et al. 2014a, 2015).

To the contrary, in coral-derived bioreactors superactivated by the hTGF-$β_3$ isoform, the adjacent muscle of the pretreated macroporous bioreactors shows *BMP-2* and *TGF-β3* upregulation. This morphogenetic wave must thus initiate a sequential chain of cellular induction, rapidly recruiting mesenchymal stem cells surrounding the implanted bioreactor, which are directly transformed into secreting osteoblasts, initiating the induction of bone formation at the periphery of the macroporous bioreactor only (Figure 3.8).

## 3.4  hTGF-$β_3$ Master Gene and Gene Product: Transfiguration of Neoplastic Masses into Bone

The rapid and robust induction of bone formation by 125 μg, but particularly by 250 μg of hTGF-$β_3$ when reconstituted with either insoluble collagenous bone matrix or coral-derived macroporous bioreactors, has shown that the hTGF-$β_3$ isoform is a most powerful soluble molecular signal that rapidly induces available progenitor stem cells at a distance of the implanted bioreactors from the surgically severed rectus abdominis muscle (Figure 3.8a and e).

The robust induction of bone formation by the hTGF-$β_3$ osteogenic device in the non-human primate *P. ursinus* has now forced a reevaluation of the mechanistic insights of the induction of bone formation in primates, including humans

(Ripamonti et al. 2014a). The hTGF-$\beta_3$ isoform does not merely initiate the induction of bone formation, but rather sets the molecular and morphological scenarios of the direct "tissue transfiguration *in vivo*" (Ripamonti 2014a), a term that now defines the molecular and morphological evidence of the rapid induction of bone formation in primate tissue only, rapidly transfiguring the striated *rectus abdominis* muscle into bone.

The morphological and molecular evidence of the rapid transfiguration of muscle tissue into bone by the hTGF-$\beta_3$ osteogenic device (Ripamonti et al. 2008, 2014a, 2015; Klar et al. 2014) has indicated to the Bone Research Laboratory a further novel, as yet totally unexplored, biological function of the hTGF-$\beta_3$ isoform, that is, the selected injections of hTGF-$\beta_3$ into malignant neoplastic primary and secondary masses to induce the rapid transfiguration of the injected masses into bone to facilitate tumoral ablation and its surgical debridement (Ripamonti 2012, 2014a, 2014b). The injection of doses of the hTGF-$\beta_3$ isoform into neoplastic masses would "osteogenize" the tumor, transfiguring all available responding cells into osteoblastic-like cells, possibly altering not only the neoplastic phenotype, but also the neoplastic genotype (Ripamonti 2012, 2014b), thus controlling differentiation so as to osteogenize secondary masses.

The rapid induction of bone formation by the hTGF-$\beta_3$ isoform has been recently shown to be formed *via* a variety of *BMP* and *TGF-$\beta$* genes profiled temporospatially expressed at selected time points controlling the complex multicellular multigene cascade of the induction of bone formation (Ripamonti et al. 2015). The data are once again challenging the *status quo* of the induction of bone formation, certainly in primates, whereby *"Bone: Formation by autoinduction"* (Urist 1965) is initiated by the expression of *BMP* genes with subsequent induction of bone formation by the secreted *BMP* gene products (Ripamonti et al. 2015). Paraphrasing the leading-edge editorial in *Cell*, "The Power of One" (2014), novel research "is exposing now the secrets of single cells and the power of single molecules." As previously stated (Ripamonti et al. 2015), while *Cell* asks the vibrant question, "But what can we learn from studying merely a single molecule?" our unique systematic studies in *P. ursinus* have indicated that in primates, and in primates only, the *TGF-$\beta$3* gene and gene product singly, yet synergistically and synchronously, set into motion the ripple-like cascade of *"Bone: Formation by autoinduction"* (Urist 1965).

To end, I copy verbatim the final conclusive statements of the classic editorial comment of M.R. Urist in "The Reality of a Nebulous Enigmatic Myth" (1968):

> Students of the problem of bone induction mindful of the many skeletal system disorders of unknown etiology, will continue with increasing vigor to explore the positive effects of this field of investigation. Julia Oppenheimer's essays (MIT Press, 1967) offer a measure of inspiration (for future research workers who need it) and a sound appraisal for the mistakes of the past.

Perhaps our studies on the hTGF-$\beta_3$ isoform in primate species may help to rectify our misunderstandings of the past on the bone induction principle and to finally appreciate the reality of the induction of bone formation in clinical contexts.

## References

BENJAMIN, L.E., HEMO, I., KESHET, E. (1998). A plasticity window for blood vessel remodelling is defined by pericyte coverage of the preformed endothelial network and is regulated by PDGF-B and VEGF. *Development* 125, 1591–98.

CAPLAN, A.I. (2008). All MSCs are pericytes? *Cell Stem Cell* 3(3), 229–30. doi: 10.1016/j.stem.2008.08.008.

CHEN, C.-W., MONTELATICI, E., CRISAN, M., CORSELLI, M., HUARD, J., LAZZARI, L., PÉAUT, B. (2009). Perivascular multi-lineage progenitor cells in human organs: Regenerative units, cytokine sources or both? *Cytokine Growth Factor Rev* 20, 429–34.

CRISAN, M., YAP, S., CASTEILLA, L., CHEN, C.W., CORSELLI, M., PARK, T.S., ANDRIOLO, G., SUN, B., ZHENG, B., ZHANG, L., NOROTTE, C., TENG, P.N., TRAAS, J., SCHUGAR, R., DEASY, B.M., BADYLAK, S., BUHRING, H.J., GIACOBINO, J.P., LAZZARI, L., HUARD, J., PÉAULT, B. (2008). A perivascular origin for mesenchymal stem cells in multiple human organs. *Cell Stem Cell* 3(3), 301–13.

GAZZERRO, E., GANGJI, V., CANALIS, E. (1998). Bone morphogenetic proteins induce the expression of noggin, which limits their activity incultured rat osteoblasts. *J Clin Invest* 102(12), 2106–14.

GROPPE, J., GREENWALD, J., WIATER, E., RODRIGUEZ-LEON, J., ECONOMIDES, A.N., KWIATKOWSKI, W., BABAN, K., AFFOLTER, M., VALE, W.W., IZPISUA BELMONTE, J.C., CHOE, S. (2003). Structural basis of BMP signaling inhibition by Noggin, a novel twelve-membered cystine knot protein. *J Bone Joint Surg Am* 85A(Suppl 3), 52–58.

HEIKINHEIMO, K. (1994). Stage-specific expression of decapentaplegic-Vg-related genes 2, 4, and 6 (bone morphogenetic proteins 2, 4, and 6) during human tooth morphogenesis. *J Dent Res* 3(3), 590–97.

HELDER, M.N., ÖZKAYNAK, E., SAMPATH, K.T., LUYTEN, F.P., LATIN, V., OPPERMANN, H., VUKICEVIC, S. (1995). Expression pattern of osteogenic protein-1 (bone morphogenetic protein-7) in human and mouse development. *J Histochem Cytochem* 43(10), 1035–44.

KINGSLEY, D.M. (1994). The TGF-β superfamily: New members, new receptors, and new genetic tests of function in different organisms. *Genes Dev* 8, 133–46.

KLAR, R.M., DUARTE, R., DIX-PEEK, T., DICKENS, C., FERRETTI, C., RIPAMONTI, U. (2013). Calcium ions and osteoclastogeneisis initiate the induction of bone formation by coral-derived macroporous constructs. *J Cell Mol Med* 17(11), 1444–57.

KLAR, R.M., DUARTE, R., DIX-PEEK, T., RIPAMONTI, U. (2014). The induction of bone formation by the recombinant human transforming growth factor-β₃. *Biomaterials* 35(9), 2773–88.

KOVACIC, J.C., BOEHM, M. (2008). Resident vascular progenitor cells: An emerging role for non-terminally differentiated vessel-resident cells in vascular biology. *Stem Cell Res* 2, 2–15.

LANDER, A.D. (2007). Morpheus unbound: Reimagining the morphogen gradient. *Cell* 128, 245–56.

LUO, G., HOFMAN, C., BRONKERS, A.L., SOHOCKTI, N., BRADELEY, A., KARSENTY, G. (1995). BMP-7 is an inducer of nephrogenesis, and is also required for eye development and skeletal patterning. *Genes Dev* 9, 2808–20. doi: 10.1101/gad.9.22.2808 1995.

MASSAGUÉ, J. (2000). How cells read TGF-beta signals. *Nat Rev Mol Cell Biol* 1(3), 169–78.

MICHAEL JONES, C., SMITH J.C. (1998). Establishment of a BMP-4 morphogen gradient by long-range inhibition. *Dev Biol* 194, 12–17.

ÖZKAYNAK, E., SCHNEGELSBERG, P.N., OPPERMANN, H. (1991). Murine osteogenic protein-1 (OP-1): High levels of mRNA in kidney. *Biochem Biophys Res Commun* 179(1), 116–23.

PIEK, E., HENDRIK-HELDIN, C., TEN DIJKE, P. (1999). Specificity, diversity, and regulation in TGF-β superfamily signalling. *FASEB J* 13, 2105–24.

REDDI, A.H. (2000). Morphogenesis and tissue engineering of bone and cartilage: Inductive signals, stem cells, and biomimetic biomaterials. *Tissue Eng* 6(4), 351–59.

RIPAMONTI, U. (1991). The morphogenesis of bone in replicas of porous hydroxyapatite obtained from conversion of calcium carbonate exoskeleton of coral. *J Bone Joint Surg Am* 73, 692–703.

RIPAMONTI, U. (2003). Osteogenic proteins of the transforming growth factor-β superfamily. In H.L. Henry and A.W. Norman (eds.), *Encyclopedia of Hormones*. Academic Press, San Diego, CA, pp. 80–86.

RIPAMONTI, U. (2004). Soluble, insoluble and geometric signals sculpt the architecture of mineralized tissues. *J Cell Mol Med* 8(2), 169–80.

RIPAMONTI, U. (2005). Bone induction by recombinant human osteogenic protein-1 (hOP-1, BMP-7) in the primate *Papio ursinus* with expression of mRNA of gene products of the TGF-β superfamily. *J Cell Mol Med* 9, 911–28.

RIPAMONTI, U. (2006). Soluble osteogenic molecular signals and the induction of bone formation. *Biomaterials* 27, 807–822.

RIPAMONTI, U. (2007). Recapitulating development: A template for periodontal tissue engineering. *Tissue Eng* 13, 51–71.

RIPAMONTI, U. (2009). Biomimetism, biomimetic matrices and the induction of bone formation. *J Cell Mol Med* 13(9B), 2953–72.

RIPAMONTI, U. (2012). Osteogenic device for inducing bone formation in clinical contexts. U.S. Patent 2012/0277879 A1, November 1.

RIPAMONTI, U. (2014a). Transfiguration of neoplastic tumoral masses into bone for superior surgical debridement. NRF Blue Skies Funding Instrument—Concept Notes, NRF Grant 93117.

RIPAMONTI, U. (2014b). Osteogenic device for inducing bone formation in clinical contexts. EP Office 06 809 015.8-1455.

RIPAMONTI, U., CROOKS, J., KHOALI, L., RODEN, L. (2009c). The induction of bone formation by coral-derived calcium carbonate/hydroxyapatite constructs. *Biomaterials* 30, 1428–39.

RIPAMONTI, U., CROOKS, J., MATSABA, T., TASKER, J. (2000) Induction of endochondral bone formation by recombinant human transforming growth factor-β$_2$ in the baboon (*Papio ursinus*). *Growth Factors* 17(4), 269–85.

RIPAMONTI, U., DIX-PEEK, T., PARAK, R., MILNER, B., DUARTE, R. (2015). Profiling bone morphogenetic proteins and transforming growth factor-βs by hTGF-β$_3$ pre-treated coral-derived macroporous constructs: The power of one. *Biomaterials* 49, 90–102. doi: 10.1016/j.biomaterials.2015.01.056.

RIPAMONTI, U., DUARTE, R., FERRETTI, C. (2014a). Re-evaluating the induction of bone formation in primates. *Biomaterials* 35(35), 9407–22.

RIPAMONTI, U., DUNEAS, N., VAN DEN HEEVER, B., BOSCH, C., CROOKS, J. (1997). Recombinant transforming growth factor-β$_1$ induces endochondral bone in the baboon and synergizes with recombinant osteogenic protein-1 (bone morphogenetic protein-7) to initiate rapid bone formation. *J Bone Miner Res* 12, 1584–595.

RIPAMONTI, U., FERRETTI, C., HELIOTIS, M. (2006). Soluble and insoluble signals and the induction of bone formation: Molecular therapeutics recapitulating developnent. *J Anat* 209, 447–68.

RIPAMONTI, U., FERRETTI, C., TEARE, J., BLANN, L. (2009a). Transforming growth factor-β isoforms and the induction of bone formation: Implications for reconstructive craniofacial surgery. *J Craniofac Surg* 20, 1544–55.

RIPAMONTI, U., HERBST N.-N., RAMOSHEBI, L.N. (2005). Bone morphogenetic proteins in craniofacial and periodontal tissue engineering: Experimental studies in the non-human primate *Papio ursinus*. *Cytokine Growth Factor Rev* 16, 357–68.

RIPAMONTI, U., KLAR, R.M., DUARTE, R., DIX-PEEK, T. (2014b). Engineering microenvironments superactivated by hTGF-$\beta_3$ reprogramming recruited differentiated pericytes into highly active secreting osteoblasts in primate striated muscles. Presented at Proceedings of the Keystone Symposia on Engineering Cell Fate and Function: Stem Cells and Reprogramming, Olympic Valley, CA, April 6–11.

RIPAMONTI, U., KLAR, R.M., RENTON, L.F., FERRETTI, C. (2010). Synergistic induction of bone formation by hOP-1 and TGF-$\beta_3$ in macroporous coral-derived hydroxyapatite constructs. *Biomaterials* 31(25), 6400–10.

RIPAMONTI, U., MA, S., CUNNINGHAM, N., YATES, L., REDDI, A.H. (1992). Initiation of bone regeneration in adult baboons by osteogenin, a bone morphogenetic protein. *Matrix* 12, 202–12.

RIPAMONTI, U., PARAK, R., PETIT, J.C. (2009b). Induction of cementogenesis and periodontal ligament regeneration by recombinant human transforming growth factor-$\beta_3$ in Matrigel with rectus abdominis responding cells. *J Periodont Res* 44(1), 81–87.

RIPAMONTI, U., PETIT, J.-C. (2009). Bone morphogenetic proteins, cementogenesis, myoblastic stem cells and the induction of periodontal tissue regeneration. *Cytokine Growth Factor Rev* 20(5–6), 489–99.

RIPAMONTI, U., RAMOSHEBI, L.N., MATSABA, T., TASKER, J., CROOKS, J., TEARE, J. (2001). Bone induction by bmps/ops and related family members in primates. *J Bone Joint Surg Am* 83A(Suppl 1, Pt 2), S116–27.

RIPAMONTI, U., RAMOSHEBI L.N., PATTON J., MATSABA T., TEARE J., RENTON L. (2004). Soluble signals and insoluble substrata: Novel molecular cues instructing the induction of bone. In E.J. Massaro and J.M. Rogers (eds.), *The Skeleton*. Humana Press, Totowa, New Jersey, pp. 217–27.

RIPAMONTI, U., RAMOSHEBI, L.N., TEARE, J., RENTON, L., FERRETTI, C. (2008). The induction of endochondral bone formation by transforming growth factor-$\beta_3$: Experimental studies in the non-human primate *Papio ursinus*. *J Cell Mol Med* 12(3), 1029–48.

RIPAMONTI, U., RODEN, L. (2010a). Induction of bone formation by transforming growth factor-$\beta_2$ in the non-human primate *Papio ursinus* and its modulation by skeletal muscle responding stem cells. *Cell Prolif* 43, 207–18.

RIPAMONTI, U., RODEN, L. (2010b). Biomimetics for the induction of bone formation. *Expert Rev Med Devices* 74(4), 469–79.

RIPAMONTI, U., RODEN, L.C., RENTON, L.F. (2012b). Osteoinductive hydroxyapatite-coated titanium implants. *Biomaterials* 33, 3813–23.

RIPAMONTI, U., TEARE, J., FERRETTI, C. (2012a). A macro-porous bioreactor superactivated by the recombinant human transforming growth factor-β3. *Front Physiol* 3, 172. doi: 10.3389/fphys.2012.00172.

RIPAMONTI, U., VAN DEN HEEVER, B., VAN WYK, J. (1993). Expression of the osteogenic phenotype in porous hydroxyapatite implanted extraskeletally in baboons. *Matrix* 13, 491–502.

SAMPATH, T.K., RASHKA, K.E., DOCTOR, J.S., TUCKER, R.F., HOFFMANN, F.M. (1993). *Drosophila* TGF-β superfamily proteins induce endochondral bone formation in mammals. *Proc Natl Acad Sci USA* 90, 6004–8.

TEN DIJKE, P., IWATA, K.K., GODDARD, C., PIELER, C., CANALIS, E., MCCARTHY, T.L., CENTRELLA, M. (1990). Recombinant transforming growth factor-β3: Biological activities and receptor-binding properties in isolated bone cells. *Mol Cell Biol* 10(9), 4473–79.

THE POWER OF ONE. (2014). Introduction. *Cell*, 157, 3.

THOMADAKIS, G., RAMOSHEBI, L.N., CROOKS, J., RUEGER, D.C., RIPAMONTI, U. (1999). Immunolocalization of bone morpho-genetic protein-2 and -3 and osteogenic protein-1 during murine tooth root morphogenesis and in other craniofacial structures. *Eur J Oral Sci* 107(5), 368–77.

TOPPER, J.N., CAI, J., QIU, Y., ANDERSON, K.R., XU, Y.Y., DEED, J.D., FEELEY, R., GIMENO, C.J., WOOLF, E.A., TAYBER, O., MAYS, G.G., SAMPSON, B.A., SCHOEN, F.J., FALB, D. (1997). Vascular MADs: Two novel MAD-related genes selectively inducible by flow in human vascular endothelium. *Proc Natl Acad Sci USA* 94(17), 9314–19.

UNIQUELY HUMAN BIOLOGY. (2014). *Cell* 157, 215.

URIST, M.R. (1965). Bone: Formation by autoinduction. *Science* 159, 893–99.

URIST, M.R. (1968). The reality of a nebulous enigmatic myth. *Clin Orthop Relat Res* 59, 3–5.

VAINIO, S., KARAVANOVA, I., JOWETT, A., THESLEFF, I. (1993). Identification of BMP-4 as a signal mediating secondary induction between epithelial and mesenchymal tissues during early tooth development. *Cell* 75(1), 45–58.

WOZNEY, J.M., ROSEN, V., CELESTE, A.J., MITSOCK, L.M., WHITTERS, M.J., KRIZ, R.W., HEWICK, R.M., WANG, E.A. (1988). Novel regulators of bone formation: Molecular clones and activities. *Science* 242, 1528–34.

ZHENG, B., CAO, B., CRISAN, M., SUN, B., LI, G., LOGAR, A., YAP, S., POLLETT, J.B., DROWLEY, L., CASSINO, T., GHARAIBEH, B., DEASY, B.M., HUARD, J., PEAULT, B. (2007). Prospective identification of myogenic endothelial cells in human skeletal muscle. *Nat Biotechnol* 25, 1025–34.

# CHAPTER FOUR

# Induction of Bone Formation by the Mammalian Transforming Growth Factor-βs

## *Molecular and Morphological Insights*

*Raquel Duarte,[1] Kurt Lightfoot,[1] and Ugo Ripamonti[2]*

[1]Department of Internal Medicine, School of Clinical Medicine, Faculty of Health Sciences, University of the Witwatersrand, Johannesburg, Johannesburg, South Africa

[2]Bone Research Laboratory, School of Oral Health Sciences, Faculty of Health Sciences, University of the Witwatersrand, Johannesburg, Johannesburg, South Africa

### 4.1 Introduction

Transforming growth factor-β (TGF-β) is a multifunctional, secreted homodimeric protein that regulates tissue homeostasis by controlling a myriad of processes, such as cellular proliferation, differentiation, apoptosis, and migration. TGF-$\beta_1$ is the prototypic member of a superfamily of more than 35 cytokines that includes the TGF-βs, the bone morphogenetic proteins (BMPs), growth and differentiation factors (GDFs), Nodal, and activins. The members of the TGF-β family play critical roles in embryonic development and adult tissue homeostasis, and any perturbations of the signaling pathways underlie developmental disorders and human disease (Massagué et al. 2000).

In mammals, three isoforms have been identified, TGF-$\beta_1$, -$\beta_2$, and -$\beta_3$, whose expression levels differ between different tissue types but have comparable signaling properties, signaling through common receptors and Smad proteins (Massagué and Weis-Garcia 1996). TGF-$\beta$ is secreted in a latent precursor form and is retained in the extracellular matrix (ECM). In response to specific signals, TGF-$\beta$ is activated proteolytically to form a homodimer consisting of two 12.5 kd polypeptides linked through a disulfide bond (Annes et al. 2003). TGF-$\beta$ signals across the plasma membrane by forming high-affinity complexes with membrane-bound serine/threonine kinase receptor (TGF-$\beta$ receptor type II, T$\beta$RII or Tgfbr2), which recruits and phosphorylates TGF-$\beta$ receptor type I (T$\beta$RI), also known as activin-receptor-like kinase 5 (ALK-5) (Shi and Massagué 2003). Phosphorylation occurs in the juxtamembrane region/GS domain, which is glycine and serine residue rich.

Intracellular signaling is initiated by T$\beta$RI through the phosphorylation of receptor-regulated Smads (R-Smads), Smad-2 and -3, at C-terminal serine residues at the SSXS motif (Shi and Massagué 2003). Access to T$\beta$RI is mediated by accessory proteins, such as the Smad anchor for receptor activation (SARA). The activated R-Smads then form a heterodimer with a common mediator Smad (Co-Smad), Smad-4, to be translocated to the nucleus to drive cell-type transcriptional activities. The transcriptional response is driven through the interaction of the Smad complex with transcription factors and their high-affinity binding to Smad cis-acting elements in the promoters of TGF-$\beta$ responsive genes. Smad partner transcription factors include members of the forkhead, zinc finger, Ets, basic helix–loop–helix, and AP1 families (Derynck et al. 1998; Ross and Hill 2008). The selection of a broad repertoire of transcription factors as partners allows the Smads to exhibit a wide range of pleiotropic effects. In addition, the Smad complexes engage co-activator and co-repressor proteins to fine-tune the transcriptional response (Figure 4.1). Both Smad-3 and Smad-4 have sequence-specific DNA binding activity despite having low affinity for DNA. In a conserved domain of the Smads, an 11-residue $\beta$-hairpin contacts the 5'-AGAC-3' cis-acting element, known as the Smad binding element (SBE) (Shi et al. 1998).

Inhibitory Smads (I-Smads) (Smad-6 and Smad-7) are a subclass of Smads that oppose the activion of R-Smads to antagonize the signaling pathway and function in a variety of ways (Hayashi et al. 1997). They compete with R-Smads for binding to type I receptors to prevent the phosphorylation of the R-Smads. They have subsequently been found to recruit E3 ubiquitin ligases, called Smad ubiquitination regulatory factors (Smurfs) (Smurf1 and Smurf2), to the TβRI to promote receptor ubiquitination and degradation, resulting in the cessation of signaling (Itoh and ten Dijke 2007).

## 4.2 Role of TGF-β in Differentiation of Osteoblasts

The TGF-β/BMP pathways are important mediators of bone formation during development and display multiple functions in the adult body. In the developing embryo, a BMP gradient is created and BMP is critical for ectodermal cell differentiation and dorsal–ventral axis determination. Any perturbations in TGF-β/BMP signal transduction result in a variety of bone diseases, such as osteoarthritis and tumor metastasis (Zhen et al. 2013; Ye et al. 2011).

The process of bone formation entails two very distinct phases: that of endochondrial ossification, during which cartilage is replaced with bone, and intramembranous ossification, during which bones are formed directly from mesenchymal cell condensations (Chen et al. 2012). The function and integrity of bone are upheld by the delicate balance between two major cell types: the osteoblasts, cells that deposit new bone, and osteoclasts, cells that resorb bone. TGF-β operates as a vital coordinator of bone modeling by orchestrating the activities of these two specialized cell types (Harada and Rodan 2003). TGF-β produced by osteoblasts is secreted in a latent inactive form to be deposited in the matrix of bone. Here, it is stored in large amounts until is activated by the acidic environment created by the osteoclasts (Oursler 1994). Following activation, TGF-β stimulates the proliferation of osteoblast precursors and matrix production. TGF-β is critical for the maintenance and expansion of mesenchymal stem cells, the osteoblast progenitors that are regulated through both autocrine and paracrine signaling mechanisms (Derynck and Akhurst 2007). Cartilage and bone are enriched with TGF-β and are the targets of TGF-β activity. During early development, fetal osteoblastic cell populations are responsive to the mitogenic cues of TGF-β, and TGF-β promotes the proliferation, differentiation, and osteoblastic commitment of osteoprogenitors. The intracellular signals generated by TGF-β are extensive and include phosphorylation of intracellular Smads; the regulation of Ras and Rho pathways; the activation of kinases, including the mitogen-activated

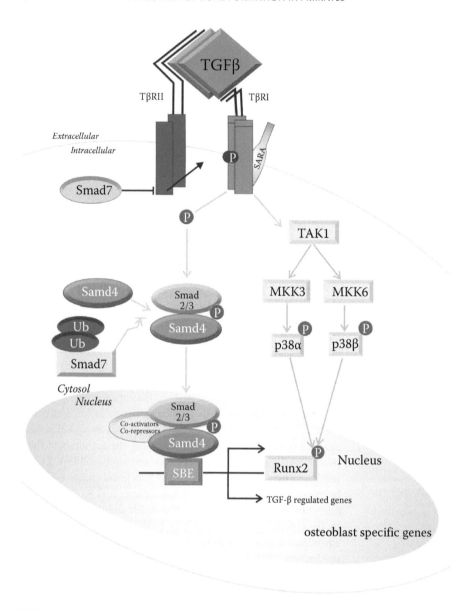

**FIGURE 4.1** The TGF-β signaling pathway in bone induction. In the canonical Smad-dependent pathway, bioactive TGF-β binds TβRII, which recruits and phosphorylates TβRI with the subsequent phosphorylation of the intracellular mediators, the R-Smads (Smad-2 and Smad-3). The activated Smads then complex with Smad-4 and translocate to the nucleus, where they interact with transcription factors, co-activators, and co-repressors to drive the expression of TGF-β-regulated genes in osteoblasts, Runx2 being one important bone-inducing factor. Inhibitory Smads (Smad-6 and Smad-7) inhibit the pathway by competing for binding of the Smads at the receptor or preventing the binding of Smad-4. Access to TβRI is mediated by accessory proteins, such as

protein kinases (MAPKs), TGF-β activated kinase-1 (TAK1), and Src; and the regulation of ion channels. TGF-β also elicits its effect through signaling cross talk with a number of other critical pathways implicated in bone formation, such as the *Wingless* (WNT), bone morphogenetic protein (BMP), fibroblast growth factor (FGF), and parathyroid (PTH) pathways.

**Canonical and Noncanonical TGF Signaling in Bone Formation: Evidence from Genetic Mouse Models**

*Canonical TGF-β Signaling Pathway* Over the past few years, many studies have focused on understanding the role of TGF-β isoforms in the induction of bone formation. The three TGF-β isoforms, together with their receptors (TβRI/ALK5 and TβRII/Tgfbr2), are important inducers of intramembranous ossification, as evidenced in a range of transgenic mouse experiments spanning the past decade and a half (summarized in Table 4.1). Null mice have been generated for each of the isoforms, but only *TGF-β1* null mice survive postnatally, albeit for a few days. Mice deficient in the TGF-$\beta_1$ isoform show a reduction in the growth of bone, bone mineralization, and elasticity (Figure 4.2) (Geiser et al. 1998). *TGF-β2* null mice display a range of developmental defects, among which are several skeletal defects (Sanford et al. 1997), while mice with targeted disruption of the *TGF-β3* gene exhibit defects in cleft palate development (Kaartinen et al. 1995). In addition, *TGF-β2* null mice showed changes in nearly every organ system and died primarily due to defects in the cardiovascular system. Phenotypic analysis of targeted disruption of the three TGF-β isoforms showed very distinct and essentially nonoverlapping phenotypes. When the same organs are affected, the underlying mechanisms differ. Defects of the cleft palate are observed in both *TGF-β2* and *TGF-β3* null mice, but the defect in *TGF-β2* knockout mice is as a result of craniofacial malformations, whereas in *TGF-β3* null mice, there is a lack of palatable shelve fusion.

To further expand the analysis of the role of the three TGF-β isoforms in development, a series of double mutants and conditional mutants (including various components of the signaling pathway) were created. Mice with a double knockout of the TGF-$\beta_2$ and -$\beta_3$ isoforms were shown to lack development of the distal rib cage (Dünker and Krieglstein 2002).

Targeted disruption of the *Tgfb2* receptor gene resulted in embryonic lethality. *Tgfbr2* null mice closely resembled the

---

**FIGURE 4.1** *Continued*

SARA. In the noncanonical or Smad-independent pathway, PTH binds and activates the PTH1R, stimulating several downstream effector pathways, including the MAPKs. (From Chen, G., et al., *Int J Biol Sci* 8(2), 272–88, 2012.)

**Table 4.1**    Bone Phenotypes in Mouse Models for TGF-β Signaling

| Gene | Transgene Detail | Phenotype | References |
|------|------------------|-----------|------------|
| **Ligand** | | | |
| *TGF-β1* | *TGF-β1*(−/−) | Postnatal death (around 4 weeks) | Geiser et al. 1998 |
| | | Decreased bone mass and elasticity | |
| *TGF-β2* | *TGF-β2*(−/−) | Perinatal death | Sanford et al. 1997 |
| | | Craniofacial defects, reduced bone size, and ossification | |
| | | Cleft palate defects | |
| | | Skeletal defects, including rotated limbs, shortened radius and ulna, sternum and rib defects, defects of the lumbar and thoracic region vertebrae that fail to fuse | |
| *TGF-β3* | *TGF-β3*(−/−) | Die within 24 h of birth | Proetzel et al. 1995 |
| | | Cleft palate defects | Kaartinen et al. 1995 |
| *TGFβRI/ALK5* | *Dermo1-Cre* | Defects in short bones | Matsunobu et al. 2009 |
| | | Ectopic cartilaginous protrusions | |
| *Tgfbr2* | *Tgfbr2*(−/−) | Embryonic lethality at E10.5 | Oshima et al. 1996 |
| | *dnTgfbr2* | Hypoplastic cartilage | Hiramatsu et al. 2011 |
| **Receptor** | | | |
| | *Col2a1-Cre* | Defects in long bone formation | Baffi et al. 2004 |
| | *Wnt1-Cre* | Defects in cell proliferation and differentiation of osteogenic cells | Iwata et al. 2010 |
| | | Defects in mandibular development | Oka et al. 2007 |
| | *Prx1-Cre* | Short limbs | Seo and Serra 2007 |
| | | Fusion of joints of the phalanges | |
| | *Col2a1-Cre* | Defects in the base of the skull | Baffi et al. 2004 |
| | | Defects in the vertebrae | |
| **Intracellular signaling intermediaries** | | | |
| *Smad-3* | *Smad-3*(−/−) | Osteopenia | Borton et al. 2001 |
| | | Reduced cortical and cancellous bone | |
| *Smad-4* | *Smad-4*(−/−) | Embryonic lethality | Sirard et al. 1998 |
| | *TTR-Cre* | Embryonic lethality at E7.5–E9.5 | |
| | | Lack of head fold and anterior embryonic structures | Li et al. 2010 |
| | *OC-Cre* | Reduced bone mineral density | Tan et al. 2007 |
| | | Decreased bone volume and bone formation rate | |

**Table 4.1 (*Continued*)**   Bone Phenotypes in Mouse Models for TGF-β Signaling

| Gene | Transgene Detail | Phenotype | References |
|------|------------------|-----------|------------|
|        | *Col2a1-Cre* | Dwarfism | Zhang et al. 2005 |
| *Smad-7* | *Prx1-Cre* | Poor cartilage formation | Iwai et al. 2008 |
| *TAK1* | *Prx1-Cre* | Cartilage defects | Gunnell et al. 2010 |
|        | *Osx-Cre* | Clavicular hypoplasia | Greenblatt et al. |
|        |            | Delayed fonatelle fusion | 2010 |

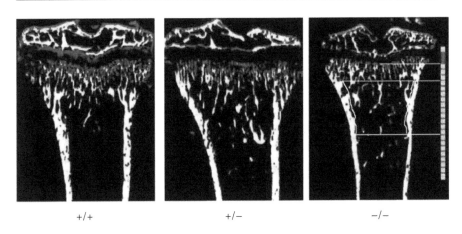

+/+                                  +/−                                  −/−

**FIGURE 4.2**   The proximal tibia of *TGF-β1* null mice. A sagittal section of proximal tibia from TGF-β₁ mice is illustrated. The micrographs (original magnification × 24) are representative of 25-day-old TGF-β₁$^{-/-}$, TGF-β₁$^{+/-}$, and wild-type (TGF-β₁$^{+/+}$) mice. The growth plate of the null mice displays reduced tissue size and a thinner growth plate. GP, growth plate; PS, primary spongiosa; SS, secondary spongiosa (Geiser et al. 1998).

*TGF-β1* null phenotype, which die early in embryonic development from defects in vasculogenesis and hematopoiesis (Oshima et al. 1996). Death of these embryos occurred too early in development to permit comparisons to the null phenotypes of *TGF-β2* and *TGF-β3*.

The capacity to investigate the role of TGF-β signaling in bone formation was facilitated through use of the *Cre/Lox* recombinase system (Sauer 1998), which permitted the development of tissue-specific conditional knockouts of key components of the TGF-β pathway. Conditional *Dermo1-Cre* knockout of TβRI in early skeletal progenitor cells resulted in mice with shorter, wider long bones and a reduction in bone collars and trabecular bone (Matsunobu et al. 2009). Mice expressing a dominant negative form of TβRII in the condensing mesenchymal cells developed hypoplastic cartilage (Hiramatsu et al. 2011).

TβRII is important in the developing axial skeleton and for maintaining the boundaries of the sclerotome (Baffi et al. 2006). *Col2a1-Cre* deletion of *Tgfbr2* resulted in mice with defects in the skull base and vertebrae, while *Prx-Cre*-driven deletion of *Tgfbr2* resulted in transgenic mice with skull vault, long bone, and joint defects (Seo and Serra 2007, 2009). These defects illustrate the important role that Tgfr2 plays in intramembranous and endochondral bone formation. Additionally, in a study by Iwata et al. (2010), *Wnt-1Cre*-targeted knockdown mice showed a loss of TβRII function leading to cranial neural crest cell (CNCC)–derived progenitor cell defects throughout intramembranous bone formation.

To define the role of TGF-β signaling in early limb development, Spagnoli and colleagues (2007) generated a mouse in which Tgfbr2 was conditionally inactivated in developing limbs. Characterization of the Tgfbr2[PRX-1KO] mice showed that they failed to form the joint interzone and therefore lacked development of interphalangeal joints (Spagnoli et al. 2007). Furthermore, the authors show that Tgfbr2 signaling was required to regulate the expression of a number of joint morphogens, including Noggin, GDF5, and Jagged, and therefore Tgfbr2 acted as the entry port for joint development.

Regulators of R-Smad activation are a group of small C-terminal domain phosphatases (SCPs) (Knockaert et al. 2006). Knockdown of SCP1/2 by RNA-mediated interference altered the sites of dephosphorylation of Smad-2 and -3, with dephosphorylation occurring preferentially at inhibitory residues of the linker region and not the normal activation C-terminal site (Sapkota et al. 2006). SCP knockdown inhibited the normal TGF-β transcriptional program.

An additional mechanism controlling R-Smad function is through activation of Smad-7, which results in degradation of the TGF-β receptor via the ubiquitin pathway (Shi and Massagué 2003). Smad-7 functions by interacting with the activated TβRI complex, preventing the binding of the R-Smads. In an in vitro cell-based study by Kavsak et al. (2000), it was demonstrated that Smad-7 forms a partnership with Smurf2, an E3 ubiquitin ligase, for the efficient degradation of the activated receptor complex. Mutants in the Smad-7 binding motif disrupted Smurf2 binding, thereby impairing their degradation function (Kavsak et al. 2000).

Smad-7 has also been shown to compete with Smad-2/3 to form a complex with Smad-4, the common mediator of both the TGF-β and BMP signaling pathways (Derynck and Zhang

2003). Through *in vitro* studies in osteoblast cell cultures, the important role of Smad-4 in osteoblast differentiation was shown to be through its interaction with a number of molecules, including the transcription factors Runx2 and Fos (Lee et al. 2000; Lai and Cheng 2002).

*Smad-3* deletion mice were viable and exhibited osteopenia with less cortical and cancellous bone (Borton et al. 2001). As *Smad-4*-knockout mice proved to be embryonic lethal to establish the in vivo role of Smad-4 in bone formation, Tan and colleagues (2007) generated transgenic mice (OC-Cre) that specifically disrupted the *Smad-4* gene in osteoblasts utilizing the Cre-loxP system (Dacquin et al. 2002). Disruption of *Smad-4* in differentiated osteoblasts gave rise to lower bone mass in young mutant mice. In addition, the mice exhibited a decrease in the number and function of osteoblasts, pointing to the important role of Smad-4 in osteoblast function. There was also an alteration in the RANKL/OP signaling pathways, resulting in a decrease in the number and activity of osteoclasts, which ultimately gave rise to greater trabecular bone volume in older mice. The observed phenotype in osteoblasts correlated to a decrease in proliferation, and this was believed to be largely due to the observed decrease in the expression of *TGFb1*. TGF-$\beta_1$ is known to increase the cell population that would give rise to osteoblasts by driving proliferation and chemotaxis (Janssens et al. 2005). Tan et al. (2007) hypothesized that downregulation of TGF-$\beta_1$ may indicate the involvement of Smad-4 in a feedback loop regulating TGF-$\beta_1$ expression in osteoblasts. This study also illustrated the important role of TGF-$\beta$ signaling in pairing bone formation and resorption in the control of bone homeostasis (Tan et al. 2007).

In *Smad-4* mutant mice, signaling via the Indian hedgehog/parathyroid hormone–related protein is decreased in the growth plate, and the culture of metatarsal bones of this mutant failed to respond to TGF-$\beta_1$ (Zhang et al. 2005). These experiments demonstrated that TGF-$\beta$–Smad-4 signals are essential for endochondral ossification by controlling chondrocyte organization in the embryonic growth plate.

***Noncanonical TGF-$\beta$ Signaling Pathways*** Non-Smad-dependent TGF-$\beta$ signaling also plays an important role in osteoblast differentiation and bone formation. In a study disrupting ALK5 function, Matsunobu et al. (2009) utilized a tamoxifen-inducible *Cre-ER*-mediated system to assess the effects of ALK5 deficiency in primary calvarial-derived cell culture. It was revealed that TGF-$\beta$ signals osteoblastic differentiation by promoting progenitor proliferation and commitment through activation of the MAPK and Smad-2/3 signaling pathways (Mastunobu et al. 2009). The association of TGF-β activation kinase 1 (TAK1), a member of the TGF-β MAP kinase kinase kinase (MAPKKK) family, with the TAK1 binding protein-1 (TAB1), triggers a kinase cascade, thereby fulfilling important TGF-β signaling intermediaries roles. They function by activating the MKK3-p38 MAPK pathway, leading to type I collagen expression. The use of double-negative TAK1 constructs illustrated that TAK1 acts to regulate steady-state MKK3 and p38 MAPK protein levels (Kim et al. 2007). Following induction by TGF-β, both the p38 MAPK and Smad signaling pathways come together to control expression of *Runx2* and regulate differentiation of mesenchymal progenitor cells (Lee et al. 2002). Activation of the ERK-MAPK pathway by TGF-$\beta_2$ activation promotes cell proliferation of osteoprogenitor cells, leading to osteoblast differentiation and the rapid expansion of calvarial bone (Lee et al. 2006).

**TGF-β Signaling Cross Talk in Bone Formation**

TGF-β controls a multitude of cellular events, and the activities of the TGF-β isoforms are influenced by many other signals. It is this relationship between TGF-β and other pathways that is responsible for the intricacy, variety, and plasticity of TGF-β function.

The interaction of TGF-β with the other pathways is complex and is highly dependable on cellular context. TGF-β is capable of communicating with other pathways at many levels, as outlined in Figure 4.3.

*TGF-β and MAPK Signaling*   The evolutionary conserved MAPKs (Erk1/2, p38/MAPKs, and JNK1/2/30) are essential regulators of a myriad of cellular functions. The MAPK pathway can be triggered by multiple extracellular stimuli to initiate a phosphorylation cascade from MAP kinase kinase kinase (MAPKKK) to MAP kinase kinase (MAPKK), and then finally MAPK. The MAPK in turn phosphorylates a wide range of proteins, including transcription factors, to effect cellular changes (reviewed in Guo and Wang 2009). The MAPKs are also under regulation of TGF-β/BMP signaling and are an important mode of non-Smad-dependent TGF-β signaling; they are discussed in Section 4.2.1.2.

Results from several studies have shown that the linker regions of the Smads present an important area for assimilating MAPK and TGF-β signaling (Matsuura et al. 2005). The linker is a flexible region of the Smads and loosely organized,

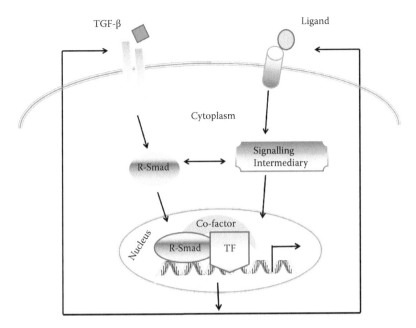

**FIGURE 4.3** Modes of TGF-β signaling cross talk. TGF-β interacts with other signaling pathways on various levels. Interactions can take place between components of the signaling pathways in the cytoplasm or the nucleus of the cell and are context dependent. Cross talk may often also occur as part of a feedback mechanism. (From Guo, X., and Wang, X.F., *Cell Res* 19(1), 71–88, 2009.)

permitting easy access to kinases. Being serine, proline, and threonine, this site also favors phosphorylation by kinases, MAPK, and glycogen synthase kinase 3-β (GSK3-β). Reports have indicated that activation of Erk or JNK by RTKs results in phosphorylation of Smad-2/3. Three residues in the linker regions of Smad-3 (Ser 203, Ser 207, and Thr 178) are the phosphorylation sites by Erk1/2 and result in translocation of Smad-3 to the nucleus with an inhibition of Smad-3 transcriptional activity (Matsuura et al. 2005). The aforementioned serine residues are also the sites of phosphorylation by other kinases, illustrating the layers of complexity that exist in signaling cross talk mechanisms.

***Wnt and TGF-β Signaling***   Wnt proteins are secreted molecules that regulate cell proliferation, differentiation, cell migration, and survival. The transcriptional co-factor β-catenin is the mediator of canonical Wnt signaling functioning by shuttling between the nucleus and cytoplasm of the cell and through it, interaction with cadherin, which is important for establishing adherence junctions between cells (Peifer and Polakis 2000).

When Wnt is absent, cytoplasmic levels of β-catenin are controlled by a complex of Axin, adenomatous polyposis coli (APC), GSK3, and casein kinase Iα (CKIα), the so-called β-catenin destruction complex. This complex is responsible for directing the phosphorylation, polyubiquitination, and degradation of β-catenin, keeping the Wnt pathway in an OFF state. When Wnt binds its receptor and co-receptor, Frizzled and LRP5/6, respectively, a signal is transduced by the intracellular mediator Dishevelled (Dvl). Axin is downregulated, GSK3 inactivated, and β-catenin stabilized. β-Catenin levels increase in the cytoplasm and are translocated to the nucleus, where they interact with the TCF/Lef (T-cell factor/lymphocyte enhancer factor) transcription factors to activate the Wnt pathway (Peifer and Polakis 2000).

Cross talk between the TGF-β and Wnt pathways has been most extensively studied, and the two pathways, linked throughout development, are able to interact at many levels at the molecular level. Each pathway is responsible for regulating the production of the other's ligand. This reciprocity is believed to be important for setting up the extracellular morphogen gradients required for modulating intricate developmental processes. In the cytoplasm, there are interactions between various components of the two pathways, which is believed to be an important system for fine-tuning the signaling process. The nucleus presents a vital component of the signaling cross with the Smad/β-catenin/Lef complex to regulate a multitude of genes, often occurring synergistically.

TGF-β collaborates with Wnt signaling in promoting differentiation of human mesenchymal stem cells along the osteoblast lineage. Synthetic β-catenin knockdown using siRNA technology increased alkaline phosphatase (ALP) activity and reversed the inhibition of sialoprotein expression in bone by TGF-$\beta_1$. TGF-β triggers β-catenin signaling through the ALK5, Smad-3, and PKA pathway and controls osteoblastogenesis through the ALK5, PKA, and JNK signaling (Zhou 2011). In general, activation of the Wnt pathway has been shown to have either a stimulatory or inhibitory effect on TGF-β action, depending on cellular context. In a series of osteoblast differentiation experiments, the Wnt pathway was shown to stabilize β-catenin and increase TCF/Lef-controlled gene expression in concert with the activity of a β-catenin-independent signaling pathway involving TCF-4 and RunX2, ultimately resulting in an increase in TβR1 gene expression (McCarthy and Centrella 2010).

In human mesenchymal stem cells (hMSCs), Smad-3-β-catenin cross talk has been observed (Jian et al. 2006). The

TGF-β and Wnt pathways cooperate to stimulate the proliferation of the hMSCs, thereby inhibiting differentiation of the cells along the osteocytic and adipocytic lineages and maintaining the self-renewal program of the progenitor cells. Stimulation of the hMSCs with TGF-β resulted in a rapid co-localization of β-catenin and Smad-3 into the nucleus, where they co-regulated a family of genes that were previously not thought of as being targets of either TGF-β or Wnt signaling, namely, Src tyrosine kinase BLK (Jian et al. 2006). What this transcriptional program of TGF-β/Wnt signaling means for stem cell renewal remains to be determined.

***TGF-β and PTH Signaling***    Mice with osteoblastic deficiency in *Tgfbr2* exhibited an increase in bone mass shown to be due to a hyperactivation of the PTH type I receptor (PTH1R). An injection of PTH or disruption of the PTH1R is able to rescue the phenotype of these mice. At a molecular level, studies have revealed that Tgfbr2 phosphorylates the cytoplasmic domain of PTH1R (Qiu et al. 2010). PTH functions in bone formation and resorption by internalizing Tgfb2 and PTH1R to attenuate downstream signaling. The cAMP response element binding (CREB) protein drives the PTH transcriptional response, and the PTH-CREB pathway acts to stimulate BMP-2 expression (Zhang et al. 2011).

***TGF-β and FGF***    FGF-2, -4, and -6 and TGF-β are inducers of osteoblast proliferation and mineralization during bone formation. FGF acts downstream of TGF-β to regulate proliferation of the CNCCs of the frontal primordium (Sasaki et al. 2006). In addition, the proliferation defect observed in the *Tgfr2* mutant mice was rescued by FGF-2. These lines of evidence demonstrate the importance of FGF signaling, arbitrated by TGF-β, in the developing frontal bone. Additionally, evidence exists for the role of the FGF/FGFR3 axis in determining the TGF-β effect on bone formation in the embryo (Mukherjee et al. 2005).

***TGF-β and BMP Signaling Cross Talk***    Several lines of evidence point toward a powerful association between TGF-β and BMP signaling in osteogenesis. Binary application of recombinant hBMP-7 with relatively low doses of hTGF-$\beta_1$ (20:1 ratio) synergized to construct large ossicles in the rectus abdominis muscle of *Papio ursinus* (Ripamonti et al. 1997). A BMP-7/TGF-$\beta_1$ binary application induced a two- to threefold increase in generated ossicles when compared to BMP-7 alone (Ripamonti et al. 1997).

TGF-$\beta_1$ strongly enhances ectopic bone formation by BMP-2 (Tachi et al. 2011). When collagen sponges treated with 50 ng of BMP-2 were implanted into the muscle tissue of mice, bone induction was observed at 7 days after implantation. However, when the BMP-2 sponges were co-treated with 50 ng of TGF-$\beta_1$, there was accelerated bone ectopic bone formation with a fivefold increase in the bone volume compared to BMP-2-only-treated sponges. There was also a corresponding twofold increase in the osteoblast number in the ectopic bone formed in the BMP-2/TGF-$\beta_1$-treated sponges (Tachi et al. 2011).

In vitro studies in human primary bone cells have shown that TGF-$\beta_1$ regulates the BMP pathway by significantly and preferentially upregulating the BMP receptor (BMPR-IB) expression through BMP-2 induction (Singhatanadgit et al. 2006). The underlying molecular mechanism is due to the augmenting effect of TGF-$\beta$ on BMP-2, which differentially activates BMPR-IIB followed by Smad-1/5/8 phosphorlyation, Dlx5 expression, and alkaline phosphatase (ALP) activity.

In a study by Wu et al. (2010), the T$\beta$Rs were shown to be necessary to mediate the osteogenic effect of BMP-9. The construction of a series of dominant negative type II receptor BMPRs reduced the BMP-9-induced osteogenic differentiation program of C3H10T1/2 stem cells (Wu et al. 2010). This study provided evidence for the strong link between the two pathways in controlling the differentiation of osteoblasts.

The TGF-$\beta$/BMP pathways are also under sophisticated regulation to sustain the accurate signaling required for the morphogenesis in tissues and organs. TGF-$\beta$/BMP activity undergoes negative regulation at the extramembranous level by ligand antagonists, intracellularly via Smurf and inhibitory Smads, and at the transcriptional level via repressor proteins and by epigenetic mechanisms.

Ligand antagonists targeting TGF-$\beta$ (Chordin and Follistatin) and the BMPs (Noggin) have been uncovered (Harland 2008; Gajos-Michniewicz et al. 2010; Krause et al. 2011). Noggin, the most characterized antagonist of bone induction pathways, regulates the activity of BMP during dorsal–ventral pattern formation, joint formation, and skeletogenesis by binding the domain normally bound by BMP-7 (Walsh et al. 2010). Dominant Noggin bone-targeted transgenic mice suffer osteopenia, a decrease in trabecular bone volume, and defective osteoblastic activity (Devlin et al. 2003; Wu et al. 2003).

Smurf1 and Smurf2 function to suppress TGF-$\beta$ signal transduction by degrading the Smads and TGF-$\beta$ and BMP receptor proteins (Datto and Wang 2005). Smurf1 degrades

Smad-1/5 and the transcriptional regulator, Runx2 (Zhao et al. 2004). Transgenic mice deficient in *Smurf1* showed an age-related increase in bone mass (Yamashita et al. 2005). Smurf2 interacts with Smad-2 and is responsible for directing the transcriptional repressor SnoN to undergo degradation (Bonni et al. 2001). Transgenic *Smurf2* mice show a decrease in articular cartilage (Wu et al. 2008). Additional proteins, such as the carboxy terminus of Hsc70-interacting protein (CHIP), are inhibitors of bone-forming TGFβ/BMP signals and act by engaging Smad-1/5 from the active R-Smad complex to undergo ubiquitination and degradation (Wang, L. et al. 2011).

Within the nucleus the transcriptional repressor proteins Ski and SnoN, together with the I-Smads, function as negative regulators by disrupting the TGF-β-stimulated active Smad–promoter complex to inhibit transcription (Stroschein et al. 1999; Zhang et al. 2007). At the molecular level, the Ski binding site on Smad-4 overlaps the binding site for phosphorylated R-Smad proteins such that binding of Ski or SnoN perturbs the interaction of Smad-4 with the R-Smad complex (Wu et al. 2002). At the DNA level, Ski and SnoN prevent Smad proteins from interacting with the transcriptional co-activator p300/CBP, an interaction that is essential for skeleton formation.

At the epigenetic level, DNA methylation is important for ALP gene expression during the osteoblast-to-osteocyte transition during osteogenesis (Delgado-Calle et al. 2011). Smad-2 induced transcription in response to TGF-β signaling involves the histone acetyltransferase p300 (Tu and Luo 2011). Mutations of lysines-19, -20, and -39 of p300 eradicated acetylation of Smad-2, thwarting nuclear accumulation of Smad-2 and preventing a TGF-β signal response (Ross et al. 2006). Furthermore, Runx2, the master transcriptional regulator of osteogenesis, is vital for controlling bone-related gene expression and is maintained by a delicate balance between acetylation and deacetylation, and ubiquitination. Signaling via BMP-2 triggers p300-mediated acetylation of Runx2, which increases its transactivation ability while at the same time inhibiting its degradation by Smurf1 (Jeon et al. 2006).

## 4.3  Role of TGF-β₃ in Bone Formation: Lessons from the Non-Human Primate *Papio ursinus*

A series of studies in the Chacma baboon (*P. ursinus*), conducted by the Bone Research Laboratory of the University of the Witwatersrand, Johannesburg, have provided profound insight into the induction of bone formation by members of the TGF-β family. In the non-human primate *P. ursinus*, the formation of bone by autoinduction was first reported when implanting engineered coral-derived carbonate, fully converted hydroxyapatite constructs (HA/CC) in the *rectus abdominus* muscle of the non-human primate (Ripamonti 1990, 1991). Since then, Ripamonti and colleagues have revealed a formerly unknown function for the three mammalian TGF-β isoforms, that of endochondral bone induction when the recombinant human isoforms are implanted into heterotopic intramuscular sties of the Chacma baboon (Ripamonti et al. 1997, 2000, 2008; Ripamonti and Roden 2010). hTGF-$\beta_3$ (125 and 250 μg) was shown to be capable of inducing the largest heterotopic ossicles reported thus far, with induction at the periphery of the implanted profile of the macroporous bioreactors used as carrier as early as 20 days after implantation (Ripamonti et al. 2008, 2012; Klar et al. 2014). This rapid induction of bone by hTGF-$\beta_3$ occurs together with the expression of several genes implicated in bone formation, namely, *BMP-2* and *-3*, *OP-1*, *TGF-β1*, and *TGF-β3*. The change in gene expression profile is accompanied by a distinct morphology, with the newly forming bone characterized by osteoblastic activity, osteoid production, angiogenesis, and capillary sprouting, the latter two events being reflected at the molecular level by increased type IV collagen expression (Ripamonti et al. 2014; Klar et al. 2014).

A working hypothesis of the Bone Research Laboratory is that the TGF-$\beta_3$-treated heterotopic implants activate the bone-forming BMP pathway, with the induction of bone formation as a secondary response consequence. Analyses of tissues produced as a result of the human TGF-$\beta_3$ isoform application have exhibited expression of *BMP-3*, *OP-1* (*BMP-7*), and *BMP-2*. To exemplify the activating role of TGF-$\beta_3$ expression on BMP pathway activation and endochondral bone induction, the inhibitory activity of Noggin was investigated in an experiment where 125 μg of recombinant human Noggin was loaded together with hTGF-$\beta_3$ into 7% hydroxyapatite coral-derived macroporous constructs. Devices were implanted into the rectus abdominus muscle of the adult baboon and the extent of bone formation and analogous gene expression patterns assayed on days 15, 60, and 90 after implantation (Klar et al. 2014).

Tissues harvested from devices loaded with 125 μg of TGF-$\beta_3$ and 125 μg of Noggin exhibited downregulation of *BMP-2* and *TGF-β* and correlated with negligible bone formation. hNoggin was observed to inhibit *BMP-2* expression on days 15 and 60, but was recovered by day 90, illustrating the temporal nature of the bone inducing gene expression changes.

Analysis of gene expression patterns underlying the bone induction cascade by hTGF-$\beta_3$ revealed significant upregulation of *Run-x2* and *Osteocalcin*, genes that are important for controlling the differentiation of mesenchymal stem cell progenitors along the osteoblastic lineage. The corresponding morphological analyses of day 15 tissue revealed an engineered microenvironment rich in differentiating osteoprogenitor cells in the surrounding muscle tissue.

More in-depth analysis of hTGF-$\beta_3$-treated macroporous devices on day 15 showed the expansion of fibrin/fibronectin rings into the macroporous spaces, structurally reorganizing the tissue for the differentiation patterning and morphogenesis underlying bone formation. These advancing rings of the extracellular matrix provide a structural component for hyperchromatic osteoblasts undergoing reprogramming from myoblastic/pericytic cells being recruited from the surrounding muscle tissue. Recent studies provide support for these findings. Fibronectin has been shown to arbitrate the cell fate of mesoderm, thereby indicating the value of the extracellular matrix in driving the signaling program responsible for mesoderm differentiation (Cheng et al. 2013). High-power magnification of these macroporous spaces in the bioreactors shows the intimate relationship of the invading differentiating cells and the structured fibrin/fibronectin rings. While the macroporous constructs display this very distinct patterning of extracellular matrix invasion at day 15, there is no bone formation yet in hTGF-$\beta_3$-treated macroporous constructs.

Of interest to note is that longitudinally implanted hTGF-$\beta_3$ devices were observed to be fused together, forming large blocks of newly formed bone. An important observation of the 125 µg TGF-$\beta_3$ plates of bone is that there is expansion of bone from the periphery of the devices into the surrounding muscle tissue, resulting in the envelopment of the constructs in the newly forming heterotopic ossicles. The induction of bone extending away and beyond the periphery of the implanted macroporous constructs is reminiscent of previous observations of the synergy observed following hOP-1/hTGF-$\beta_1$ co-treatment at a 20:1 ratio of weight of the implanted devices (Ripamonti et al. 1997).

An important feature of bone formation of the synergistic activation of the binary application (TGF-$\beta_1$/OP-1) and the 125 µg of hTGF-$\beta_3$ treatment is the apparent development of a maturational gradient (Ripamonti et al. 1997, 2014) of tissue organization within the developing ossicle, with morphogenesis occurring toward the periphery of the ossicle and taking the form of corticalization of a mineralized shell of bone to resemble trabeculae of bone encased by an osteoid seam. In contrast to the newly forming bone at the periphery is the presence of voids toward the center of ossicles when collagenous bone matrix is used as a carrier. This is suggestive of a constriction on ossicle growth due to limited vascularization toward the center, and hence limited cellular invasion.

Further evidence for a growth factor maturation gradient is provided when 250 µg of hTGF-$\beta_3$-treated devices is implanted in the *rectus abdominus* muscle of the Chacma baboon. There is prominent induction of bone up to 3–4 cm away from the implanted macroporous carrier. This extensive bone formation, with accompanying trabeculation, forms at the periphery, with very little induction within the device itself. The observed superactivation by the 125 µg of hTGF-$\beta_3$ isoform has been hypothesized to be a result of enhanced recruitment of responding progenitor cells at the implantation site and resulting rapid induction of bone (Ripamonti et al. 2012, 2014). Alternatively, the prominent osteoinduction could be due to a diffusion gradient that has been established, setting in motion a sequence of cellular induction events resulting in the induction of bone at the site of optimal morphogen concentration.

Recently, studies have provided evidence for the role of latent TGF-$\beta$ as a molecular sensor that, following any cellular perturbations to the extracellular matrix, due to either wound repair, tissue injury, or inflammation, is responsible for the release of active TGF-$\beta$. Wan and colleagues (2012) have shown that in response to tissue injury, there is an activation of TGF-$\beta_1$, with the active TGF-$\beta_1$ acting to stimulate the migration of MSCs to the sites of injury in order to promote remodeling of tissue and vascular repair (Wan et al. 2012). Additionally, higher levels of the active TGF-$\beta_1$ in the peripheral blood are responsible for recruiting MSCs from the bone marrow into the peripheral blood supply. Thus, following tissue injury, TGF-$\beta_1$ may have a dual role, acting as a systemic factor to mobilize MSC and a local factor to promote MCP1 production and cause the migration of MSCs to the site of injury. An important source of active TGF-$\beta_1$ is the vascular matrix, since it is known that the ECM acts to concentrate latent TGF-$\beta_1$ at sites of tissue repair and provides a reservoir for potential activators. In addition, earlier findings from in vitro studies indicate that for MSCs, the "microenvironmental niche" is critical in defining cell fate responses to molecular signals (Gregory et al. 2005).

Ripamonti et al. (2015) provided an expression profile for key members of the TGF-$\beta$ superfamily in the induction of bone when either hTGF-$\beta_3$-treated or untreated bioreactors were implanted in the *rectus abdominis* muscle of *P. ursinus*. The important role of the surrounding microenvironment was elegantly demonstrated by molecular profiling of the various TGF-$\beta$ and BMP isoforms in the muscle immediately adjacent to the implanted macroporous bioreactors (Ripamonti et al. 2015). Relative changes in gene expression of the *BMP* genes, *BMP-2*, *BMP-3*, *BMP-4*, and *BMP-6*, and the three *TGF-$\beta$* isoforms were assessed using quantitative reverse-transcriptase polymerase chain reaction (qRT-PCR), and results are presented in Figure 4.4. At 60 days after implantation in the untreated macroporous devices, there was minimal expression of *BMP-2*, *BMP-4*, *BMP-6*, and *BMP-7* in the device itself, and the expression of these genes in the adjacent muscle was very similar to that observed in the devices. Of the BMPs, *BMP-3* showed the most prominent upregulation. hTGF-$\beta_3$ treatment resulted in the upregulation of *BMP-3* and *BMP-4*, while the expression of *BMP-2* and *BMP-6* remained the same as in the untreated device. An interesting observation was the effect of hTGF-$\beta_3$ treatment on the surrounding muscle tissue, which exhibited increased *BMP-3* and *BMP-7* expression. Also in this in vivo system, *BMP-7* was decreased in the device following 250 µg of TGF-$\beta$ treatment, with greater *BMP-7* expression levels evident in the adjacent muscle tissue.

The expression profile of the *TGF-$\beta$* isoform genes is characterized by a differential expression pattern in the bone-inducing in vivo model. In the untreated device at 60 days after implantation, there is upregulation of the *TGF-$\beta1$* and *TGF-$\beta3$* expression with a concomitant decrease in the *TGF-$\beta2$* expression. The expression profile following implantation of 250 µg of treated hTGF-$\beta_3$ devices shows a marked increase in the expression of *TGF-$\beta1$*, while *TGF-$\beta2$* expression levels remain unchanged from the untreated sample, and there is a decrease in the *TGF-$\beta3$* expression. Of great interest is that there is expression of all three *TGF-$\beta$* isoforms in the tissue immediately adjacent to all implanted bioreactors.

In a recent *in vitro* study to characterize the effect of TGF-$\beta$ isoforms on osteoblast cells, Sefat et al. (2014) found that treatments with the different TGF-$\beta$ isoforms elicited different effects, with TGF-$\beta_3$ responsible for the greatest increase in cell proliferation and migration, TGF-$\beta_2$ causing the greatest amount of cell hypertrophy, and TGF-$\beta_1$ having the least effect on cell size, proliferation, and migration. This study is

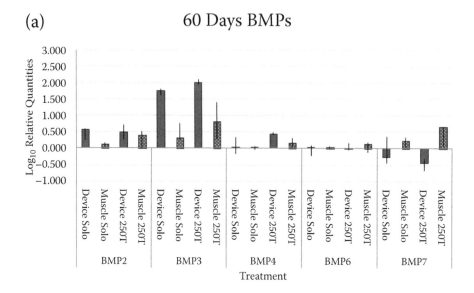

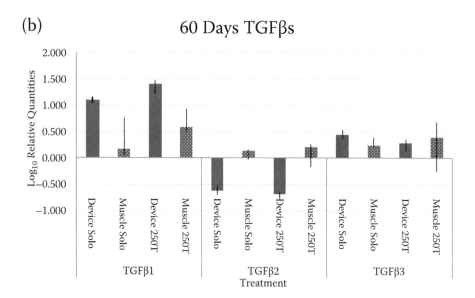

**FIGURE 4.4** Relative expression of *BMP* (a) and *TGF-β* (b) isoforms at 90 days after heterotopic implantation in the rectus abdominis of *P. ursinus*. Seven percent HA/CC macroporous devices, either untreated (solo) or treated with 250 μg of hTGF-β₃ (250 T). By using a quantitative RT-PCR approach, the relative gene expression of the *TGF-B/BMP* isoforms in the device and in the muscle tissue immediately adjacent to the device is shown. The bars are indicative of medians and lines in the upper and lower quartile ranges ($p^* < .05$, $** < .01$) (Ripamonti et al. 2015).

illustrative of the fact that osteoblast activity in response to TGF-β isoforms is dependent on the differentiation state of the cells and the cell types constituting the microenvironment.

Morphologically, the 250 μg treated hTGF-β$_3$ devices exhibited extensive induction of bone formation, albeit at the periphery of the implanted macroporus bioreactor with lack of bone formation in the internal spaces. The bone formed was characterized by trabeculae covered by osteoid seams and populated by adjoining osteoblastic cells opposing the highly vascular mesenchymal tissue (Figure 4.5).

## 4.4  Perspectives

Cumulative studies over the past two decades have demonstrated the critical role for TGF-β superfamily members in bone formation. Newer technologies such as the generation of conditional knockout mice have rapidly advanced understanding of the process. The signaling pathway through which TGF-β isoforms control cellular events has been well established with each stage in the signaling process—from binding of the ligand to the receptor, to signaling *via* intracellular intermediaries, and finally to how these signals impact gene expression, being finely dissected. The discovery of signaling cross talk between TGF-β signaling and other biologically important pathways illustrates their important regulatory functions in osteoblast differentiation and the complexity of the bone induction process. The in vivo model provided by the Chacma baboon has played a pivotal role in understanding the morphological and molecular events underlying bone induction, and thus assisting in providing answers to the fundamental question of tissue engineering: What is it that underlies pattern formation? The untreated or morphogen-treated coral-derived macroporous bioreactors offers a microenvironmental niche enabling the transformation, differentiation, and migration of pericytic stem cells from the highly vascular adjacent muscle tissue. This dynamic microenvironment then provides for a cascade of molecular signaling events that orchestrate the osteogenic program, resulting in the induction of bone formation at the site of the implanted bioreactors. Continued research will reveal how the TGF-β signaling pathway is woven into the bone induction mechanism. It is envisioned that the combination of morphological and molecular approaches in the in vivo bone model provided by the Chacma baboon will greatly assist in unraveling the complex mechanisms underlying the induction

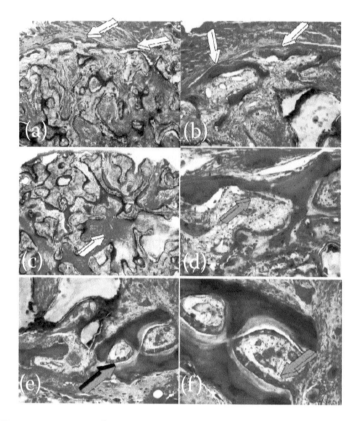

**FIGURE 4.5** Tissue morphogenesis by coral-derived macroporous bioreactors pre-loaded with 250 μg of recombinant human TGF-$\beta_3$ harvested 60 days after implantation in the rectus abdominis muscle of *P. ursinus*. (a, b) The induction of bone is initiated only at the very periphery of the bioreactors (white arrows), with a distinctive lack of bone within the central areas (white arrow in c). (d–f) Newly formed bone at the periphery of the bioreactor. The newly formed mineralized bone is surfaced by osteoid seams populated by adjoining osteoblasts (light blue arrow) facing the invading sprouting capillaries. (From Ripamonti, U., et al., *Biomaterials* 49, 90–102, 2015. doi: 10.1016/j.biomaterials.2015.01.056.)

of bone formation by the three mammalian TGF-β isoforms in primates species only.

## References

ANNES, J.P., MUNGER, J.S., RIFKIN, D.B. (2003). Making sense of latent TGFbeta activation. *J Cell Sci* 116(Pt 2), 217–24.

BAFFI, M.O., MORAN, M.A., SERRA, R. (2006). Tgfbr2 regulates the maintenance of boundaries in the axial skeleton. *Dev Biol* 296(2), 363–74.

BAFFI, M.O., SLATTERY, E., SOHN, P., MOSES, H.L., CHYTIL, A., SERRA, R. (2004). Conditional deletion of the TGF-beta type II receptor in Col2a expressing cells results in defects in the axial skeleton without alterations in chondrocyte differentiation or embryonic development of long bones. *Dev Biol* 276(1), 124–42.

BONNI, S., WANG, H.R., CAUSING, C.G., KAVSAK, P., STROSCHEIN, S.L., LUO, K., WRANA, J.L. (2001). TGF-beta induces assembly of a Smad2-Smurf2 ubiquitin ligase complex that targets SnoN for degradation. *Nat Cell Biol* 3(6), 587–95.

BORTON, A.J., FREDERICK, J.P., DATTO, M.B., WANG, X.F., WEINSTEIN, R.S. (2001). The loss of Smad3 results in a lower rate of bone formation and osteopenia through dysregulation of osteoblast differentiation and apoptosis. *J Bone Miner Res* 16(10), 1754–64.

CHEN, G., DENG, C., LI, Y.P. (2012). TGF-$\beta$ and BMP signaling in osteoblast differentiation and bone formation. *Int J Biol Sci* 8(2), 272–88.

CHENG, P., ANDERSEN, P., HASSEL, D., KAYNAK, B.L., LIMPHONG, P., JUERGENSEN, L., KWON, C., SRIVASTAVA, D. (2013). Fibronectin mediates mesendodermal cell fate decisions. *Development* 140(12), 2587–96.

DACQUIN, R., STARBUCK, M., SCHINKE, T., KARSENTY, G. (2002). Mouse alpha1(I)-collagen promoter is the best known promoter to drive efficient Cre recombinase expression in osteoblast. *Dev Dyn* 224(2), 245–51.

DATTO, M., WANG, X.F. (2005). Ubiquitin-mediated degradation a mechanism for fine-tuning TGF-beta signaling. *Cell* 121(1), 2–4.

DELGADO-CALLE, J., SAÑUDO, C., SÁNCHEZ-VERDE, L., GARCÍA-RENEDO, R.J., AROZAMENA, J., RIANCHO, J.A. (2011). Epigenetic regulation of alkaline phosphatase in human cells of the osteoblastic lineage. *Bone* 49(4), 830–38.

DERYNCK, R., AKHURST, R.J. (2007). Differentiation plasticity regulated by TGF-beta family proteins in development and disease. *Nat Cell Biol* 9(9), 1000–4.

DERYNCK, R., ZHANG, Y.E. (2003). Smad-dependent and Smad-independent pathways in TGF-beta family signaling. *Nature* 425(6958), 577–84.

DERYNCK, R., ZHANG, Y., FENG, X.H. (1998). Smads: Transcriptional activators of TGF-beta responses. *Cell* 95(6), 737–40.

DEVLIN, R.D., DU, Z., PEREIRA, R.C., KIMBLE, R.B., ECONOMIDES, A.N., JORGETTI, V., CANALIS, E. (2003). Skeletal overexpression of noggin results in osteopenia and reduced bone formation. *Endocrinology* 144(5), 1972–78.

DÜNKER, N., KRIEGLSTEIN, K. (2002). Tgfbeta2 –/– Tgfbeta3 –/– double knockout mice display severe midline fusion defects and early embryonic lethality. *Anat Embryol* (Berl) 206(1–2), 73–83.

GAJOS-MICHNIEWICZ, A., PIASTOWSKA, A.W., RUSSELL, J.A., OCHEDALSKI, T. (2010). Follistatin as a potent regulator of bone metabolism. *Biomarkers* 15(7), 563–74.

GEISER, A.G., ZENG, Q.Q., SATO, M., HELVERING, L.M., HIRANO, T., TURNER, C.H. (1998). Decreased bone mass and bone elasticity in mice lacking the transforming growth factor-beta1 gene. *Bone* 23(2), 87–93.

GREENBLATT, M.B., SHIM, J.H., ZOU, W., SITARA, D., SCHWEITZER, M., HU, D., LOTINUN, S., SANO, Y., BARON, R., PARK, J.M., ARTHUR, S., XIE, M., SCHNEIDER, M.D., ZHAI, B., GYGI, S., DAVIS, R., GLIMCHER, L.H. (2010). The p38 MAPK pathway is essential for skeletogenesis and bone homeostasis in mice. *J Clin Invest* 120(7), 2457–73.

GREGORY, C.A., YLOSTALO, J., PROCKOP, D.J. (2005). Adult bone marrow stem/progenitor cells (MSCs) are preconditioned by microenvironmental "niches" in culture: A two-stage hypothesis for regulation of MSC fate. *Sci STKE* 294, pe37.

GUNNELL, L.M., JONASON, J.H., LOISELLE, A.E., KOHN, A., SCHWARZ, E.M., HILTON, M.J., O'KEEFE, R.J. (2010). TAK1 regulates cartilage and joint development via the MAPK and BMP signaling pathways. *J Bone Miner Res* 25(8), 1784–97.

GUO, X., WANG, X.F. (2009). Signaling cross-talk between TGF-beta/BMP and other pathways. *Cell Res* 19(1), 71–88.

HARADA, S., RODAN, G.A. (2003). Control of osteoblast function and regulation of bone mass. *Nature* 423(6937), 349–55.

HARLAND, R.M. (2008). A protein scaffold plays matchmaker for chordin. *Cell* 134(5), 718–19.

HAYASHI, H., ABDOLLAH, S., QIU, Y., CAI, J., XU, Y.Y., GRINNELL, B.W., RICHARDSON, M.A., TOPPER, J.N., GIMBRONE, M.A., JR., WRANA, J.L., FALB, D. (1997). The MAD-related protein Smad7 associates with the TGFbeta receptor and functions as an antagonist of TGFbeta signaling. *Cell* 89(7), 1165–73.

HIRAMATSU, K., IWAI, T., YOSHIKAWA, H., TSUMAKI, N. (2011). Expression of dominant negative TGF-β receptors inhibits cartilage formation in conditional transgenic mice. *J Bone Miner Metab* 29(4), 493–500.

ITOH, S., TEN DIJKE, P. (2007). Negative regulation of TGF-beta receptor/Smad signal transduction. *Curr Opin Cell Biol* 19(2), 176–84.

IWAI, T., MURAI, J., YOSHIKAWA, H., TSUMAKI, N. (2008). Smad7 inhibits chondrocyte differentiation at multiple steps during endochondral bone formation and down-regulates p38 MAPK pathways. *J Biol Chem* 283(40), 27154–64.

IWATA, J., HOSOKAWA, R., SANCHEZ-LARA, P.A., URATA, M., SLAVKIN, H., CHAI, Y. (2010). Transforming growth factor-beta regulates basal transcriptional regulatory machinery to control cell proliferation and differentiation in cranial neural crest-derived osteoprogenitor cells. *J Biol Chem* 285(7), 4975–82.

JANSSENS, K., TEN DIJKE, P., JANSSENS, S., VAN HUL, W. (2005). Transforming growth factor-beta1 to the bone. *Endocr Rev* 26(6), 743–74.

JEON, E.J., LEE, K.Y., CHOI, N.S., LEE, M.H., KIM, H.N., JIN, Y.H., RYOO, H.M., CHOI, J.Y., YOSHIDA, M., NISHINO, N., OH, B.C., LEE, K.S., LEE, Y.H., BAE, S.C. (2006). Bone morphogenetic protein-2 stimulates Runx2 acetylation. *J Biol Chem* 281(24), 16502–11.

JIAN, H., SHEN, X., LIU, I., SEMENOV, M., HE, X., WANG, X.F. (2006). Smad3-dependent nuclear translocation of beta-catenin is required for TGF-beta1-induced proliferation of bone marrow-derived adult human mesenchymal stem cells. *Genes Dev* 20(6), 666–74.

KAARTINEN, V., VONCKEN, J.W., SHULER, C., WARBURTON, D., BU, D., HEISTERKAMP, N., GROFFEN, J. (1995). Abnormal lung development and cleft palate in mice lacking TGF-beta 3 indicates defects of epithelial-mesenchymal interaction. *Nat Genet* 11(4), 415–21.

KAVSAK, P., RASMUSSEN, R.K., CAUSING, C.G., BONNI, S., ZHU, H., THOMSEN, G.H., WRANA, J.L. (2000). Smad7 binds to Smurf2 to form an E3 ubiquitin ligase that targets the TGF beta receptor for degradation. *Mol Cel*, 6(6), 1365–75.

KIM, S.I., KWAK, J.H., ZACHARIAH, M., HE, Y., WANG, L., CHOI, M.E. (2007). TGF-beta-activated kinase 1 and TAK1-binding protein 1 cooperate to mediate TGF-beta1-induced MKK3-p38 MAPK activation and stimulation of type I collagen. *Am J Physiol Renal Physiol* 292(5):F1471–78.

KLAR, R.M., DUARTE, R., DIX-PEEK, T., RIPAMONTI, U. (2014). The induction of bone formation by the recombinant human transforming growth factor-$\beta_3$. *Biomaterials* 35(9), 2773–88.

KNOCKAERT, M., SAPKOTA, G., ALARCÓN, C., MASSAGUÉ, J., BRIVANLOU, A.H. (2006). Unique players in the BMP pathway: Small C-terminal domain phosphatases dephosphorylate Smad1 to attenuate BMP signaling. *Proc Natl Acad Sci USA* 103(32), 11940–45.

KRAUSE, C., GUZMAN, A., KNAUS, P. (2011). Noggin. *Int J Biochem Cell Biol* 43(4), 478–81.

LAI, C.F., CHENG, S.L. (2002). Signal transductions induced by bone morphogenetic protein-2 and transforming growth factor-beta in normal human osteoblastic cells. *J Biol Chem* 277(18), 15514–22.

LEE, K.S., HONG, S.H., BAE, S.C. (2002). Both the Smad and p38 MAPK pathways play a crucial role in Runx2 expression following induction by transforming growth factor-beta and bone morphogenetic protein. *Oncogene* 21(47), 7156–63.

LEE, K.S., KIM, H.J., LI, Q.L., CHI, X.Z., UETA, C., KOMORI, T., WOZNEY, J.M., KIM, E.G., CHOI, J.Y., RYOO, H.M., BAE, S.C. (2000). Runx2 is a common target of transforming growth factor beta1 and bone morphogenetic protein 2, and cooperation between Runx2 and Smad5 induces osteoblast-specific gene expression in the pluripotent mesenchymal precursor cell line C2C12. *Mol Cell Biol* 20(23), 8783–92.

LEE, S.W., CHOI, K.Y., CHO, J.Y., JUNG, S.H., SONG, K.B., PARK, E.K., CHOI, J.Y., SHIN, H.I., KIM, S.Y., WOO, K.M., BAEK, J.H., NAM, S.H., KIM, Y.J., KIM, H.J., RYOO, H.M. (2006). TGF-beta2 stimulates cranial suture closure through activation of the Erk-MAPK pathway. *J Cell Biochem* 98(4), 981–91.

LI, C., LI, Y.P., FU, X.Y., DENG, C.X. (2010). Anterior visceral endoderm SMAD4 signaling specifies anterior embryonic patterning and head induction in mice. *Int J Biol Sci* 6(6), 569–83.

MASSAGUÉ, J., BLAIN, S.W., LO, R.S. (2000). TGFbeta signaling in growth control, cancer, and heritable disorders. *Cell* 103(2), 295–309.

MASSAGUÉ, J., WEIS-GARCIA, F. (1996). Serine/threonine kinase receptors: Mediators of transforming growth factor beta family signals. *Cancer Surv* 27, 41–64.

MATSUNOBU, T., TORIGOE, K., ISHIKAWA, M., DE VEGA, S., KULKARNI, A.B., IWAMOTO, Y., YAMADA, Y. (2009). Critical roles of the TGF-beta type I receptor ALK5 in perichondrial formation and function, cartilage integrity, and osteoblast differentiation during growth plate development. *Dev Biol* 332(2), 325–38.

MATSUURA, I., WANG, G., HE, D., LIU, F. (2005). Identification and characterization of ERK MAP kinase phosphorylation sites in Smad3. *Biochemistry* 44(37), 12546–53.

MCCARTHY, T.L., CENTRELLA, M. (2010). Novel links among Wnt and TGF-beta signaling and Runx2. *Mol Endocrinol* 24(3), 587–97.

MUKHERJEE, A., DONG, S.S., CLEMENS, T., ALVAREZ, J., SERRA, R. (2005). Co-ordination of TGF-beta and FGF signaling pathways in bone organ cultures. *Mech Dev* 122(4), 557–71.

OKA, K., OKA, S., SASAKI, T., ITO, Y., BRINGAS, P., JR., NONAKA, K., CHAI, Y. (2007). The role of TGF-beta signaling in regulating chondrogenesis and osteogenesis during mandibular development. *Dev Biol* 303(1), 391–404.

OSHIMA, M., OSHIMA, H., TAKETO, M.M. (1996). TGF-beta receptor type II deficiency results in defects of yolk sac hematopoiesis and vasculogenesis. *Dev Biol* 179(1), 297–302.

OURSLER, M.J. (1994). Osteoclast synthesis and secretion and activation of latent transforming growth factor beta. *J Bone Miner Res* (4), 443–52.

PEIFER, M., POLAKIS, P. (2000). Wnt signaling in oncogenesis and embryogenesis: A look outside the nucleus. *Science* 287(5458), 1606–9.

QIU, T., WU, X., ZHANG, F., CLEMENS, T.L., WAN, M., CAO, X. (2010). TGF-beta type II receptor phosphorylates PTH receptor to integrate bone remodelling signaling. *Nat Cell Biol* 12(3), 224–34.

RIPAMONTI, U. (1990). Inductive bone matrix and porous hydroxy-apatite composites in rodents and non-human primates. In T. Yamamuro, J. Wilson-Hench, and L.L. Hench (eds.), *Handbook of Bioactive Ceramics*, Vol. II: *Calcium Phosphate and Hydroxylapatite Ceramics*. Boca Raton, FL: CRC Press, pp. 245–53.

RIPAMONTI, U. (1991). The morphogenesis of bone in replicas of porous hydroxyapatite obtained from conversion of calcium carbonate exoskeletons of coral. *J Bone Joint Surg Am* 73, 692–703.

RIPAMONTI, U., CROOKS, J., MATSABA, T., TASKER, J. (2000). Induction of endochondral bone formation by recombinant human transforming growth $\beta_2$ in the baboon (*Papio ursinus*). *Growth Factors* 17(4), 269–85.

RIPAMONTI, U., DIX-PEEK, T., PARAK, R., MILNER, B., DUARTE, R. (2015). Profiling bone morphogenetic proteins and transforming growth factor-$\beta$s by hTGF-$\beta_3$ pre-treated coral-derived macroporous constructs: The power of one. *Biomaterials* 49, 90–102. doi: 10.1016/j.biomaterials.2015.01.056.

RIPAMONTI, U., DUARTE, R., FERRETTI, C. (2014). Re-evaluating the induction of bone formation in primates. *Biomaterials* 35(35), 9407–22.

RIPAMONTI, U., DUNEAS, N., VAN DEN HEEVER, B., BOSCH, C., CROOKS, J. (1997). Recombinant transforming growth factor-beta1 induces endochondral bone in the baboon and synergizes with recombinant osteogenic protein-1 (bone morphogenetic protein-7) to initiate rapid bone formation. *J Bone Miner Res* 12(10), 1584–95.

RIPAMONTI, U., RAMOSHEBI, L.N., TEARE, J., RENTON, L., FERRETTI, C. (2008). The induction of endochondral bone formation by transforming growth factor-$\beta_3$: Experimental studies in the non-human primate *Papio ursinus*. *J Cell Mol Med* 12(3), 1029–48.

RIPAMONTI, U., RODEN, L.C. (2010). Induction of bone formation by transforming growth factor-$\beta_2$ in the non-human primate *Papio ursinus* and its modulation by skeletal muscle responding stem cells. *Cell Prolif* 43(3), 207–18.

RIPAMONTI, U., TEARE, J., FERRETTI, C.A. (2012). Macroporous bioreactor super activated by the recombinant human transforming growth factor-$\beta_3$. *Front Physiol* 3, 172.

ROSS, S., CHEUNG, E., PETRAKIS, T.G., HOWELL, M., KRAUS, W.L., HILL, C.S. (2006). Smads orchestrate specific histone modifications and chromatin remodeling to activate transcription. *EMBO J* 25(19), 4490–502.

ROSS, S., HILL, C.S. (2008). How the Smads regulate transcription. *Int J Biochem Cell Biol* 40(3), 383–408.

SANFORD, L.P., ORMSBY, I., GITTENBERGER-DE GROOT, A.C., SARIOLA, H., FRIEDMAN, R., BOIVIN, G.P., CARDELL, E.L., DOETSCHMAN, T. (1997). TGFbeta2 knockout mice have multiple developmental defects that are non-overlapping with other TGFbeta knockout phenotypes. *Development* 124(13), 2659–70.

SAPKOTA, G., KNOCKAERT, M., ALARCÓN, C., MONTALVO, E., BRIVANLOU, A.H., MASSAGUÉ, J. (2006). Dephosphorylation of the linker regions of Smad1 and Smad2/3 by small C-terminal domain phosphatases has distinct outcomes for bone morphogenetic protein and transforming growth factor-beta pathways. *J Biol Chem* 281(52), 40412–19.

SASAKI, T., ITO, Y., BRINGAS, P., JR., CHOU, S., URATA, M.M., SLAVKIN, H., CHAI, Y. (2006). TGFbeta-mediated FGF signaling is crucial for regulating cranial neural crest cell proliferation during frontal bone development. *Development* 133(2), 371–81.

SAUER, B. (1998). Inducible gene targeting in mice using the Cre/lox system. *Methods* 14(4), 381–92.

SEFAT, F., DENYER, M.C., YOUSEFFI, M. (2014). Effects of different transforming growth factor beta (TGF-β) isomers on wound closure of bone cell monolayers. *Cytokine* 69(1), 75–86.

SEO, H.S., SERRA, R. (2007). Deletion of Tgfbr2 in Prx1-cre expressing mesenchyme results in defects in development of the long bones and joints. *Dev Biol* 310(2), 304–16.

SEO, H.S., SERRA, R. (2009). Tgfbr2 is required for development of the skull vault. *Dev Biol* 334(2), 481–90.

SHI, Y., MASSAGUÉ, J. (2003). Mechanisms of TGF-beta signaling from cell membrane to the nucleus. *Cell* 113(6), 685–700.

SHI, Y., WANG, Y.F., JAYARAMAN, L., YANG, H., MASSAGUÉ, J., PAVLETICH, N.P. (1998). Crystal structure of a Smad MH1 domain bound to DNA: Insights on DNA binding in TGF-beta signaling. *Cell* 94(5), 585–94.

SINGHATANADGIT, W., SALIH, V., OLSEN, I. (2006). Up-regulation of bone morphogenetic protein receptor IB by growth factors enhances BMP-2-induced human bone cell functions. *J Cell Physiol* 209(3), 912–22.

SIRARD, C., DE LA POMPA, J.L., ELIA, A., ITIE, A., MIRTSOS, C., CHEUNG, A., HAHN, S., WAKEHAM, A., SCHWARTZ, L., KERN, S.E., ROSSANT, J., MAK, T.W. (1998). The tumor suppressor gene Smad4/Dpc4 is required for gastrulation and later for anterior development of the mouse embryo. *Genes Dev* 12(1), 107–19.

SPAGNOLI, A., O'REAR, L., CHANDLER, R.L., GRANERO-MOLTO, F., MORTLOCK, D.P., GORSKA, A.E., WEIS, J.A., LONGOBARDI, L., CHYTIL, A., SHIMER, K., MOSES, H.L. (2007). TGF-beta signaling is essential for joint morphogenesis. *J Cell Biol* 177(6), 1105–17.

STROSCHEIN, S.L., WANG, W., ZHOU, S., ZHOU, Q., LUO, K. (1999). Negative feedback regulation of TGF-beta signaling by the SnoN oncoprotein. *Science* 286(5440), 771–74.

TACHI, K., TAKAMI, M., SATO, H., MOCHIZUKI, A., ZHAO, B., MIYAMOTO, Y., TSUKASAKI, H., INOUE, T., SHINTANI, S., KOIKE, T., HONDA, Y., SUZUKI, O., BABA, K.,

KAMIJO, R. (2011). Enhancement of bone morphogenetic protein-2-induced ectopic bone formation by transforming growth factor-β1. *Tissue Eng* 17A(5–6), 597–606.

TAN, X., WENG, T., ZHANG, J., WANG, J., LI, W., WAN, H., LAN, Y., CHENG, X., HOU, N., LIU, H., DING, J., LIN, F., YANG, R., GAO, X., CHEN, D., YANG, X. (2007). Smad4 is required for maintaining normal murine postnatal bone homeostasis. *J Cell Sci* 120(Pt 13), 2162–70.

TU, A.W., LUO, K. (2007). Acetylation of Smad2 by the co-activator p300 regulates activin and transforming growth factor beta response. *J Biol Chem* 282(29), 21187–96.

WALSH, D.W., GODSON, C., BRAZIL, D.P., MARTIN, F. (2010). Extracellular BMP-antagonist regulation in development and disease: Tied up in knots. *Trends Cell Biol* 20(5), 244–56.

WAN, M., LI, C., ZHEN, G., JIAO, K., HE, W., JIA, X., WANG, W., SHI, C., XING, Q., CHEN, Y.F., JAN DE BEUR, S., YU, B., CAO, X. (2012). Injury-activated transforming growth factor β controls mobilization of mesenchymal stem cells for tissue remodeling. *Stem Cells* 30(11), 2498–511.

WANG, L., LIU, Y.T., HAO, R., CHEN, L., CHANG, Z., WANG, H.R., WANG, Z.X., WU, J.W. (2011). Molecular mechanism of the negative regulation of Smad1/5 protein by carboxyl terminus of Hsc70-interacting protein (CHIP). *J Biol Chem* 286(18), 15883–94.

WU, J.W., KRAWITZ, A.R., CHAI, J., LI, W., ZHANG, F., LUO, K., SHI, Y. (2002). Structural mechanism of Smad4 recognition by the nuclear oncoprotein Ski: Insights on Ski-mediated repression of TGF-beta signaling. *Cell* 111(3), 357–67.

WU, N., ZHAO, Y., YIN, Y., ZHANG, Y., LUO, J. (2010). Identification and analysis of type II TGF-β receptors in BMP-9-induced osteogenic differentiation of C3H10T1/2 mesenchymal stem cells. *Acta Biochim Biophys Sin* (Shanghai) 42(10), 699–708.

WU, Q., KIM, K.O., SAMPSON, E.R., CHEN, D., AWAD, H., O'BRIEN, T., PUZAS, J.E., DRISSI, H., SCHWARZ, E.M., O'KEEFE, R.J., ZUSCIK, M.J., ROSIER, R.N. (2008). Induction of an osteoarthritis-like phenotype and degradation of phosphorylated Smad3 by Smurf2 in transgenic mice. *Arthritis Rheum* 58(10), 3132–44.

WU, X.B., LI, Y., SCHNEIDER, A., YU, W., RAJENDREN, G., IQBAL, J., YAMAMOTO, M., ALAM, M., BRUNET, L.J., BLAIR, H.C., ZAIDI, M., ABE, E. (2003). Impaired osteoblastic differentiation, reduced bone formation, and severe osteoporosis in noggin-overexpressing mice. *J Clin Invest* 112(6), 924–34.

YAMASHITA, M., YING, S.X., ZHANG, G.M., LI, C., CHENG, S.Y., DENG, C.X., ZHANG, Y.E. (2005). Ubiquitin ligase Smurf1 controls osteoblast activity and bone homeostasis by targeting MEKK2 for degradation. *Cell* 121(1), 101–13.

YE, L., MASON, M.D., JIANG, W.G. (2011). Bone morphogenetic protein and bone metastasis, implication and therapeutic potential. *Front Biosci* (Landmark ed.), 16, 865–97.

ZHANG, J., TAN, X., LI, W., WANG, Y., WANG, J., CHENG, X., YANG, X. (2005). Smad4 is required for the normal organization of the cartilage growth plate. *Dev Biol* 284(2), 311–22.

ZHANG, R., EDWARDS, J.R., KO, S.Y., DONG, S., LIU, H., OYAJOBI, B.O., PAPASIAN, C., DENG, H.W., ZHAO, M. (2011). Transcriptional regulation of BMP2 expression by the PTH-CREB signaling pathway in osteoblasts. *PLoS One* 6(6), e20780.

ZHANG, S., FEI, T., ZHANG, L., ZHANG, R., CHEN, F., NING, Y., HAN, Y., FENG, X.H., MENG, A., CHEN, Y.G. (2007). Smad7 antagonizes transforming growth factor beta signaling in the nucleus by interfering with functional Smad-DNA complex formation. *Mol Cell Biol* 27(12), 4488–99.

ZHAO, M., QIAO, M., HARRIS, S.E., OYAJOBI, B.O., MUNDY, G.R., CHEN, D. (2004). Smurf1 inhibits osteoblast differentiation and bone formation in vitro and in vivo. *J Biol Chem* 279(13), 12854–59.

ZHEN, G., WEN, C., JIA, X., LI, Y., CRANE, J.L., MEARS, S.C., ASKIN, F.B., FRASSICA, F.J., CHANG, W., YAO, J., CARRINO, J.A., COSGAREA, A., ARTEMOV, D., CHEN, Q., ZHAO, Z., ZHOU, X., RILEY, L., SPONSELLER, P., WAN, M., LU, W.W., CAO, X. (2013). Inhibition of TGF-β signaling in mesenchymal stem cells of subchondral bone attenuates osteoarthritis. *Nat Med* 19(6), 704–12.

ZHOU, S. (2011). TGF-β regulates β-catenin signaling and osteoblast differentiation in human mesenchymal stem cells. *J Cell Biochem* 112(6), 1651–60.

# CHAPTER FIVE

# Regeneration of Mandibular Defects in Non-Human Primates by the Transforming Growth Factor-β₃ and Translational Research in Clinical Contexts

*Ugo Ripamonti[1] and Carlo Ferretti[1,2]*

[1]Bone Research Laboratory, School of Oral Health Sciences, Faculty of Health Sciences, University of the Witwatersrand, Johannesburg, Parktown, South Africa

[2]Division of Maxillofacial and Oral Surgery, School of Oral Health Sciences, Chris Hani Baragwanath Hospital, University of the Witwatersrand, Johannesburg, Johannesburg, South Africa

## 5.1 Tissue Induction and Regeneration of Craniomandibulofacial Defects

Reconstruction of large craniofacial defects in humans remains a grand challenge for skeletal reconstructionists, tissue engineers, and molecular biologist alike. In context, mandibular reconstruction has been and still is a challenging endeavor, despite major biological and experimental surgical advances that have hitherto resulted in the rapid generation of novel molecular and biological data that have engineered the emergence of tissue biology and, with it, tissue engineering and regenerative medicine at large (Sampath and Reddi 1981; Reddi 1994, 2000; Ripamonti 2006; Ripamonti et al. 2004, 2014a).

This unprecedented gathering of molecular and experimental results had prematurely convinced molecular biologists and skeletal reconstructionists alike that functional and morphological reconstruction of craniomandibulofacial defects was finally at hand (Ripamonti et al. 2008a). This was particularly true after convincing preclinical experimental results gathered in a variety of animal models that showed the realistic induction of bone formation across several animal species, including non-human primates, in multiple appendicular and craniofacial sites (Cook et al. 1995; Ripamonti et al. 1996, 2000, 2006, 2007, 2012, 2014a; Ripamonti 2005; Ripamonti and Klar 2010).

This theoretical potential, biologically and molecularly dissected by recent morphological and molecular experimentation (Ripamonti et al. 2014a, 2015; Klar et al. 2014), has not been translated to acceptable results in clinical contexts. Clinical trials of craniofacial and orthopedic applications, such as mandibular reconstruction and sinus-lift operations, have indicated that supraphysiological doses of single human recombinant morphogenetic proteins are needed to induce often unacceptable tissue induction while incurring significant costs without equivalence to autogenous bone grafts (Garrison et al. 2007; Mussano et al. 2007; Ripamonti et al. 2006, 2007, 2009, 2014a; Ferretti et al. 2010). The clinical experiences in craniofacial human defects treated with hBMP-2 and hOP-1 have yielded unpredictable results requiring several tens of milligrams to achieve barely perceivable clinically significant osteoinduction or failing completely (Boyne et al. 2005; Herford and Boyne 2000).

Restoring anatomical function of complex disfiguring craniofacial defects and anomalies remains a grand unsolved challenge. Those of us who have not suffered the outrage of facial deformity visited upon patients as either a developmental misfortune or the scourge of disease or violence can only imagine the effects thereof. Loss of facial features not only denies patients the most basic human functions but also robs them of a sense of identity, with all the associated mental anguish (Ripamonti and Klar 2010; Ripamonti et al. 2012).

In the 1990s, the Bone Research Laboratory of the university, together with the Division of Oral and Maxillofacial Surgery, propelled the bone induction principle by translating results in calvarial defects implanted in *Papio ursinus* with highly purified, naturally derived osteogenic protein fractions into clinical contexts to restore tissue induction and morphogenesis in mandibular defects of human patients. Defects were implanted with either highly purified, naturally derived bovine osteogenic fractions or autogenous bone grafts in a cohort of patients under the *aegis* of the University of the Witwatersrand, Johannesburg, the Committee for Research on Human Subjects (Ripamonti and Ferretti 2002; Ferretti and Ripamonti 2002).

Dehydrated diaphyseal bovine bone matrix was demineralized in five volumes of 0.5 N hydrochloric acid at room temperature. Demineralized bone matrix was dissociatively extracted in 8 M urea, 50 mM Tris, and 100 mM NaCl (pH 7.4) with protease inhibitors (Ripamonti et al. 1992; Sampath and Reddi 1981, 1983; Sampath et al. 1987). Crude extracts were exchanged with 6 M urea, 50 mM Tris (pH 7.4), and 150 mM NaCl by ultrafiltration (Ripamonti et al. 1992). Purification was by sequential chromatography on heparin–Sepharose and hydroxyapatite–Ultrogel affinity and adsorption chromatography columns, washed and eluted as described (Ripamonti et al. 1992). Protein fractions with biological activity in the rodent subcutaneous assay were concentrated and exchanged with 4 M Gdn-HCl and 50 mM Tris (pH 7.4) and loaded onto tandem Sephacryl S-200 high-resolution gel filtration columns, equilibrated and eluted as described (Ripamonti et al. 1992, 2000). Proteins were combined with human demineralized bone matrix sieved to particle size 75–420 μm; aliquots of 0.8 mg of bovine osteogenic fractions were combined to 1 g of human matrix, lyophilized in vacuum, and packed into sealed 50 ml Nunc tubes sterilized by gamma irradiation (Ripamonti and Ferretti 2002; Ferretti and Ripamonti 2002).

Selected patients with mandibular defects were implanted with either highly purified osteogenic fractions or autogenous bone grafts (Figure 5.1) (Ripamonti and Ferretti 2002; Ferretti and Ripamonti 2002). A trephine biopsy was performed 90 days postimplantation. When successful, histological examination showed that the implanted osteogenic device exhibited mineralized bone trabeculae with copious osteoid seams (Figure 5.1f and g). Bone deposition directly onto nonvital matrix provided unequivocal evidence of osteoinduction (Figure 5.1g). Of the seven patients grafted with autogenous bone grafts, five had histological evidence of osteoinduction (Ferretti and Ripamonti 2002). Though osteogenic protein fractions had highly active osteogenesis compared to autogenous bone grafts, four treated patients failed to initiate the inductive cascade as evaluated histologically in the area of the biopsies; harvested tissues from these patients showed persistence of inactive demineralized

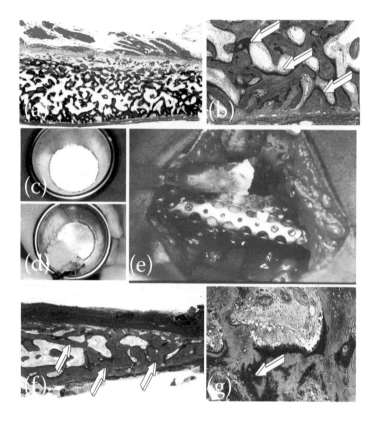

**FIGURE 5.1**   Translational research from the benchtop and from preclinical results in the Chacma baboon (*Papio ursinus*) (a, b) (Ripamonti et al. 1992) to craniomandibulo-facial defects in clinical contexts (e–g) (Ripamonti and Ferretti 2002; Ferretti and Ripamonti 2002). (a) Induction of calvarial regeneration by highly purified, naturally derived osteogenic protein fractions purified greater than 50,000-fold after sequential chromatography on hydroxyapatite–Ultrogel and heparin–Sepharose adsorption and affinity chromatography, respectively. Protein eluates exchanged to 4 M guanidinium hydrochloride were sieved by gel filtration chromatography on tandem S-200 Sephacryl columns (Ripamonti et al. 1992; Ripamonti and Ferretti 2002). (a) 280 µg of osteogenic fractions per gram of insoluble collagenous matrix as carrier induces calvarial regeneration by day 30. (b) High-power view showing newly formed mineralized bone covered by osteoid seams (arrows) populated by contiguous osteoblasts. (c, d) Lyophilized human demineralized bone matrix, combined with highly purified osteogenic protein fractions, is prepared with aliquots of sterile distilled water for implantation and adaptability within the mandibular defect (e). (f) Biopsy specimen 90 days after mandibular healing and implant incorporation showing newly mineralized bone covered by osteoid seams. (g) Osteoid seams populated by contiguous osteoblasts surfacing newly formed mineralized bone enveloping remnants of matrix carrier delivering the biological activity of the recombined osteogenic proteins.

bone matrix embedded in a dense avascular fibrous stroma (Ferretti and Ripamonti 2002).

The greatest challenge of regenerative medicine and bone tissue engineering is to translate into clinical contexts what has been so dramatically discovered in preclinical animal models, as well as on the laboratory bench (Reddi 2000; Ripamonti 2006; Ripamonti et al. 2007, 2014a). The analysis of molecular signals in solution has facilitated our understanding of the molecular mechanisms of differentiation, development, and morphogenesis of vertebrate tissues and organs (Reddi 1994, 1997, 2000, 2005). This has led to great interdisciplinary scientific challenges that have set the rules that fashion the architectural replacement of tissues and organs, and the manufacture of new tissues for the functional restoration and replacement of body parts lost to disease, trauma, and neoplastic pathologies (Reddi 2000; Ripamonti 2006), with its associated biological and functional failures (Williams 2006).

The rationale for enunciating the rules of regenerative medicine and tissue engineering is the polyhedral multifunctional biomimetism and biological activities of the extracellular matrix. The extracellular matrix is the scaffold that binds and releases soluble molecular signals, sequesters them in the right conformation to ligands on responding cells, and guides extracellular matrix deposition along guided geometrical and molecular cues (Reddi 2000; Ripamonti 2004, 2006, 2009).

Reconstruction of large craniofacial defects in humans requires the harvesting of autogenous bone grafts from a distant donor site, most often the iliac crest with associated harvest-related morbidity (Habal 1994). Additional limitations are the finite volume of available bone from donor sites and adapting the donor bone to fit the shape of the recipient defect, the final challenge to autogenous bone grafting in clinical contexts.

Hip donor sites yield large amounts of vascularized corticocancellous donor bone, particularly suited for mandibular reconstruction. In addition, the normal contour of the hip provides a near-replica contour of the mandible. The harvest of the iliac crest's graft also yields a significant amount of cancellous bone, which is excellent for encouraging ingrowth of new bone largely *via* osteoconduction (Ripamonti et al. 2008a).

The use of megadoses of recombinant hBMPs required so far in clinical osteoinduction is perhaps the most severe operational limitation of translational osteoinduction in clinical contexts. Radiographic images of hBMP-2- and hOP-1-treated human defects do not often convincingly show bone regeneration by induction across the entirety of the treated defects (Boyne et al.

2005; Herford and Boyne 2000). Corticalization and remodeling of several treated mandibular defects by hBMP-2 or hOP1 are often missing, with uninspiring radiographic evidence of tissue regeneration. In contrast, autogenous bone grafts, even when transplanted in massive mandibular defects in human patients, show unequivocal radiographic evidence of bone formation by induction across the defects (Figure 5.2).

The extent of regeneration of the avulsed segmental defect has led to the important emerging concept of *clinically significant osteoinduction*, that is, the quality and quantity of regenerated bone to be adequately identified radiographically as normal bone, in both radio-opacity and trabecular architecture (Ferretti et al. 2010; Ripamonti et al. 2014a) (Figure 5.2). The Bone Research Laboratory has raised on a few occasions the critical issue of clinically significant osteoinduction when using autogenous bone grafts, even when transplanted in massive mandibular defects (Ferretti et al. 2010; Ripamonti et al. 2014a) (Figure 5.2h), *versus* the inadequate healing by hBMP-2 and hOP-1 in comparatively smaller defects. The emerging concept of clinically significant osteoinduction and the limited, uninspiring, and inadequate healing by massive doses of recombinant hBMP-2 and hOP-1 have been raised on at least three occasions during the last International Conferences on Bone Morphogenetic Proteins (Ripamonti and Ferretti 2010; Ripamonti 2012; Ripamonti et al. 2014b). The Bone Research Laboratory has further stated, to the general indifference of the audience, that molecular biologists and skeletal reconstructionists alike need to go back to the basics to further understand the mechanisms of tissue induction by autogenous bone grafts *versus* the induction of bone by recombinant human bone morphogenetic proteins.

In this context, the Bone Research Laboratory is of the scientific opinion that the induction of bone formation by autogenous bone grafts is more than the commonly listed biological reasons, that is, viable autogenous osteoblasts and viable morcellated autogenous bone matrix with a viable supramolecular assembly of soluble autogenous morphogenetic signals (Ripamonti et al. 2014a). A concerted molecular and tissue biology effort should now be devoted to further mechanistically dissect the molecular biology of autogenous bone graft incorporation that could be successfully translated in clinical contexts when using recombinant human proteins. The need for alternatives to recombinant human bone morphogenetic proteins is now felt more acutely after reported complications and performance failure in clinical contexts (Carragee et al. 2011a, 2011b; Fauber 2011; Ripamonti et al. 2012, 2014a).

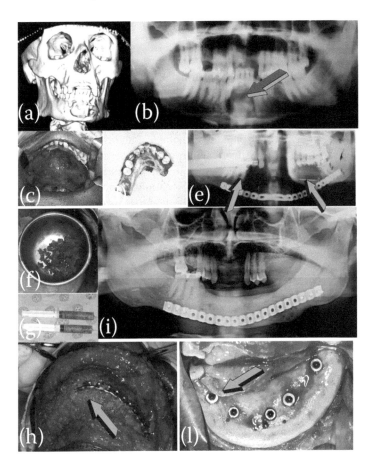

**FIGURE 5.2** Induction of bone formation in large craniomandibulofacial defects in clinical contexts by morcellated autogenous bone grafts harvested from the donor iliac crest (Ripamonti et al. 2014a). (a–d) Mandibular pathology, glandular odontogenic cyst (c), requires mandibulectomy (c, d) of the affected mandible, resulting in a severe mandibular defect (arrows in e). Harvested iliac crest bone is morcellated (f) and packed into 20 ml syringes (g) ejected along the vast mandibular defect (arrow in h). (i) Radiographic analysis 6 months after iliac crest grafting shows the quantity and quality of the regenerate, adequately to be identified radiographically as normal bone, in both radio-opacity and trabecular architecture (Ferretti et al. 2010; Ripamonti and Ferretti 2010; Ripamonti 2012). (l) Intraoperative image highlighting the quality and quantity of the regenerated bone highly suitable for the insertion of several titanium fixtures for prosthetic rehabilitation.

The above, however, will require not only a change of the use of a recombinant morphogen, member of the TGF-β supergene family, but also a thoroughly different commitment to drastically change not just the morphogen, but also the understanding of the induction of bone formation. The ancient Greek language used the word *metanoeo* (μετάνοια) to define not just merely a change to approach a problem, but a commitment or a "transformative change"; similarly, our scientific understanding must now biologically accept the new paradigm of the induction of bone formation at least in primates, where the $TGF$-$\beta_3$ gene and gene product set into motion the induction of bone formation by profiling several $BMP$ genes, ultimately initiating the rapid induction of bone formation in heterotopic *rectus abdominis* and in orthotopic mandibular sites (Ripamonti et al. 2014a, 2015). This transformative change needs now to be a fundamental change in our understanding of the induction of bone formation from preclinical settings in *P. ursinus* to clinical contexts in *Homo sapiens* (Ripamonti et al. 2014a).

## 5.2  Rapid Induction of Bone Formation by the hTGF-$\beta_3$ Osteogenic Device in Mandibular Defects of the Chacma Baboon (*P. ursinus*)

The rapid induction of mineralized bone by the hTGF-$\beta_3$ isoform, together with $TGF$-$\beta 1$, $TGF$-$\beta 3$, $BMP$-$3$, $OP$-$1$, and $BMP$-$2$ expression, hypercellular osteoblastic activity, osteoid synthesis with thick osteoid seams populated by contiguous osteoblasts, angiogenesis, and capillary sprouting with increased type IV collagen expression, when implanted in intramuscular heterotopic sites of the Chacma baboon (*P. ursinus*), is the novel molecular and morphological basis of the induction of bone formation (Ripamonti et al. 2008b, 2014a, 2014b, 2015).

hTGF-$\beta_3$ combined with coral-derived macroporous bioreactors induces prominent bone formation when implanted in the *rectus abdominis* muscle of *P. ursinus* (Ripamonti et al. 2014a; Klar et al. 2014). Of note, tissue specimens harvested on day 15 show the induction and differentiation of fibrin/fibronectin rings expanding within the macroporous spaces, structurally organizing tissue patterning and morphogenesis (Ripamonti et al. 2014a, 2014c; Klar et al. 2014). Of note, the advancing fibrin/fibronectin rings provide structural anchorage to hyperchromatic cells interpreted as differentiating osteoblasts reprogrammed by the hTGF-$\beta_3$ isoform from invading myoblastic/pericytic differentiated cells (Ripamonti et al. 2014a, 2014c). $RUNX$-$2$ and *Osteocalcin* expression are significantly upregulated in hTGF-$\beta_3$-treated bioreactors on day 15, supporting the morphological observation of invading cells differentiating into the osteoblastic phenotype with hypercellular osteoblastic activity and extracellular matrix secretion (Ripamonti et al. 2014a, 2014c).

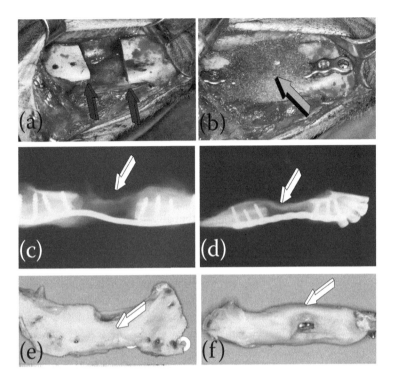

**FIGURE 5.3**    Induction of mandibular regeneration in *Papio ursinus* after implantation of 125 μg doses of recombinant human transforming growth factor-$\beta_3$ (hTGF-$\beta_3$) combined with allogeneic insoluble collagenous bone matrix as carrier, 125 μg of hTGF-$\beta_3$ per gram of carrier. (a, b) Creation of the full-thickness segmental defect, 2.5 cm in length, surgical insertion of a titanium plate, and application of the osteogenic device (b) across the defect (arrow) and over the margins of the prepared defect. (c, d) Radiographic details of the implanted defects 30 days postimplantation showing mineralization of the lingual plates (arrows). (e, f) Clinical images of harvested and operated hemimandibles showing reconstruction of the mandibular profile with some buccal expansion (arrows).

Because of the prominent induction of bone formation by the hTGF-$\beta_3$ isoform, experiments were then designed to test the efficacy of the recombinant morphogen in full-thickness segmental mandibular defects surgically prepared in *P. ursinus* (Figure 5.3).

In a first set of experiments, 125 μg of hTGF-$\beta_3$ per gram of allogeneic insoluble collagenous bone matrix implanted in full-thickness segmental mandibular defects of *P. ursinus* induced significant and unprecedented *restitutio ad integrum* with mineralization of the newly induced cortical plates as early as 30 days after implantation (Figure 5.3). Implantation of hTGF-$\beta_3$ additionally restored the mandibular profile with expansion of the newly formed buccal plate (Figure 5.3e and f).

In previous studies, our laboratories have shown that binary application of recombinant human osteogenic protein-1 (hOP-1) with hTGF-$\beta_1$ at a ratio of 20:1 yielded the most rapid induction of mineralized ossicles in the *rectus abdominis* muscle of *P. ursinus*, hOP-1 synergizing with relatively low doses of hTGF-$\beta_1$ to induce large corticalized ossicles as early as 15 days after heterotopic implantation (Ripamonti et al. 1997). The synergistic induction of bone formation has been replicated using coral-derived macroporous bioreactors as a delivery system for the synergistic induction of bone formation by binary applications of hOP-1 and hTGF-$\beta_3$, yielding massive heterotopic ossicles surrounding the implanted coral-derived constructs (Ripamonti et al. 2010).

Binary application of hOP-1 and hTGF-$\beta_3$ at a ratio of 20:1, recombined with allogeneic insoluble collagenous bone matrix, yielded prominent osteogenesis in treated mandibular defects by day 30, with significant expansion of the newly formed buccal plates, with radiographic evidence of mineralization of both buccal and lingual plates (Figure 5.4). The expansion of the corticalized buccal plates and mineralization of newly formed bone with rapid cellular recruitment for the prominent induction of bone by the synergistic induction of bone formation are primarily discussed in Chapters 2 and 6.

The prominent induction of bone formation by the recombinant morphogen by day 30 in both heterotopic *rectus abdominis* and orthotopic mandibular sites in the non-human primate *P. ursinus* has prompted translational research in clinical contexts by implanting doses of 125 µg of hTGF-$\beta_3$ per gram of human demineralized bone matrix into a large mandibular defect of a human patient. Though ossification and restoration of the mandibular defect were not in any way comparable to results obtained in *P. ursinus*, bone formed across the defect (Bone Research Laboratory 2014, unpublished data). After evaluating macroscopically and histologically the substantial induction of bone formation as initiated by 250 µg of hTGF-$\beta_3$ when combined to macroporous coral-derived bioreactors (Ripamonti et al. 2012), the prominent induction of bone formation by 250 µg of hTGF-$\beta_3$ was translated into clinical contexts to treat a large mandibular defect in a pediatric patient (Ripamonti and Ferretti 2012).

Of note, in 250 µg of hTGF-$\beta_3$-treated macroporous bioreactors there is rapid expression of signaling morphogens at the periphery of the implanted constructs, with significant cell

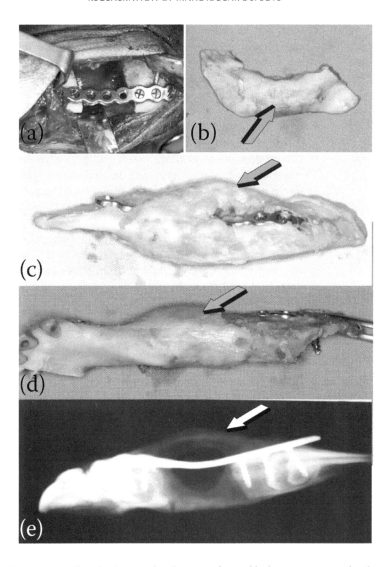

**FIGURE 5.4** Rapid and substantial induction of mandibular regeneration by the synergistic induction of bone formation after implantation of binary applications of 2.5 mg of recombinant human osteogenic protein-1 (hOP-1) with 125 μg of recombinant human transforming growth factor-$\beta_3$ (hTGF-$\beta_3$) in *Papio ursinus*. (a) Surgical preparation of the mandibular full-thickness segmental defect and insertion of the titanium plate. (b) Restitutio ad integrum of the mandibular clinical profile as early as 30 days postimplantation, without any signs of the mandibular defect (arrow). (c, d) Mandibular tissue induction with expansion of the regenerates (arrows) following extensive chemochinesis/chemotaxis of responding progenitor stem cells at the periphery of the implanted synergistic osteogenic device, with (e) prominent mineralization of both expanded buccal and lingual mineralized plates by radiographic analyses on day 30 postimplantation.

induction and differentiation within the adjacent surrounding striated *rectus abdominis* muscle. The combined molecular and morphological data indicate that there is expression of morphogenetic proteins of the TGF-β supergene family enveloping the implanted macroporous specimens, with the induction of bone formation at the periphery of the implanted bioreactor only (Ripamonti et al. 2015).

Long-term follow-up of the human case treated with the 250 μg dose of the hTGF-β₃ osteogenic device has shown the reconstruction of the large human mandibular defect with tissue induction and regeneration of the coronoid process and the ramus of the avulsed mandible (Figure 5.5) (Bone Research Laboratory 2014, unpublished data).

Continuous experimentation in both non-human and human primates is now needed to start controlled clinical trials to confirm that the implanted 250 μg of hTGF-β₃ per gram of human demineralized bone matrix is the novel dose required to set into motion tissue induction and regeneration in clinical contexts. Paraphrasing Collins's title in *Science Translational Medicine* (2011), the time is now right to reengineer the induction of bone formation by translating the hTGF-β₃ osteogenic device in clinical contexts.

The rapid induction of bone formation by the hTGF-β₃ isoform is *via* a variety of profiled *BMP* and *TGF-*β genes temporospatially expressed at selected time points controlling the complex multicellular multigene cascade of the induction of bone formation (Ripamonti et al. 2015). The data once again are challenging the *status quo* of the induction of bone formation in primates (Ripamonti et al. 2014a, 2015). As we have recently stated, paraphrasing the leading-edge editorial in *Cell*, "The Power of One" (2014), "the seminal biological discoveries made in the first 20 years of *Cell* were largely derived from pooling together large amounts of material reflecting an extrapolation from population averages rather than an understanding of the dynamics of individual players, exposing the secrets of single cells and the power of single molecules." While *Cell* asks the vibrant question "But what could one learn from studying merely a single molecule?" our systematic studies in *P. ursinus*, once again paraphrasing the *Cell* editorial, the "era of one has just begun" (The Power of One, 2014), indicate that in primates, and in primates only, the *TGF-*β₃ gene and gene product singly, yet synergistically and synchronously, set into motion the ripple-like cascade of "*Bone: Formation by autoinduction*" (Urist 1965).

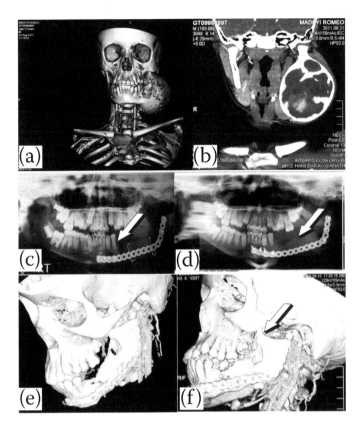

**FIGURE 5.5** Translational research in clinical contexts from the benchtop to preclinical results in the Chacma baboon (*Papio ursinus*) after the evaluation of 250 µg of recombinant human transforming growth factor-$\beta_3$ (hTGF-$\beta_3$) combined with coral-derived macroporous bioreactors implanted in the rectus *abdominis muscle* of *P. ursinus* (Ripamonti et al. 2012, 2014a). (a, b) Severe craniomandibulofacial pathology affecting the left hemimandible of a human patient. The neoplastic condition requires hemimandibulectomy, as shown in (c), from the mandibular symphysis (arrow). (d) Radiographic image of the implanted hemimandible 15 days after packaging of the hTGF-$\beta_3$ osteogenic device, 250 µg/g of gamma-irradiated human demineralized bone matrix. Arrow in (a) indicates tissue morphogenesis by day 15 within the implanted construct. Six months later (e, f), it culminates in the regeneration of the body and ramus of the newly formed mandible, with restoration of the coronoid process (arrow in f).

## References

BOYNE, P.J., LILLY, L.C., MARX, R.E., MOY, P.K., NEVINS, M., SPAGNOLI, D.B., TRIPPLET, R.G. (2005). De novo bone induction by recombinant human bone morphogenetic protein-2 (rhBMP-2) in maxillary sinus floor augmentation. *J Oral Maxillofac Surg* 63, 1693–707.

CARRAGEE, E.J., GHANAYEM, A.J., WEINER, B.K., ROTHMAN, D.J., BONO, C.M. (2011a). A challenge to integrity in spine publications: Years of living dangerously with the promotion of bone growth factors. *Spine* 11, 463–68.

CARRAGEE, E.J., HURWITZ, E.L., WEINER, B.K. (2011b). A critical review of recombinant human bone morphogenetic protein-2 trials in spinal surgery: Emerging safety concerns and lessons learned. *Spine* 11, 471–91.

COLLINS, F.S. (2011). Reengineering translational science: The time is right. *Sci Transl Med* 3(90), 90cm17. doi: 10.1126/scitranslmed.3002747.

COOK, S.D., WOLFE, M.W., SALKELD, S.L., RUGER, D.C. (1995). Effect of recombinant human osteogenic protein-1 on healing of segmental defects in non-human primates. *J Bone Joint Surg Am* 77A, 734–50.

FAUBER, J. (2011). Complications from spinal product omitted from articles: Co-authors received royalties from manufacturer. *Milwaukee Journal Sentinel*. http://www.jsonline.com/news/health/122553058.html.

FERRETTI, C., RIPAMONTI, U. (2002). Human segmental mandibular defects treated with naturally-derived bone morphogenetic proteins. *J Craniofac Surg* 13, 434–44.

FERRETTI, C., RIPAMONTI, U., TSIRIDIS, E., KERAWALA, C.J., MANTALARIS, A., HELIOTIS, M. (2010). Osteoinduction: Translating preclinical promise into clinical reality. *Br J Oral Maxillofac Surg* 48, 536–39.

GARRISON, K.R., DONELL, S., RYDER, J., SHEMILT, I., MUGFORD, M., HARVEY, I., SONG, F. (2007). Clinical effectiveness of bone morphogenetic proteins in the non-healing of fractures and spinal fusion: A systematic review. *Health Technol Assess* 11, 1–150, iii–iv.

HABAL, M.B. (2004). Bone grafting in craniofacial surgery. *Clin Plast Surg* 21, 349–63.

HERFORD, A.S., BOYNE, P.J. (2000). Reconstruction of mandibular continuity defects with bone morphogenetic protein-2 (rhBMP-2). *J Oral Maxillofac Surg* 66, 616–24. doi: 10.1016/j.joms.2007.11.02.

KLAR, R.M., DUARTE, R., DIX-PEEK, T., RIPAMONTI, U. (2014). The induction of bone formation by the recombinant human transforming growth factor-$\beta_3$. *Biomaterials* 35(9), 2773–88.

MUSSANO, F., CICCONE, G., CECCARELLI, M., BALDI, I., BASSI, F. (2007). Bone morphogenetic proteins and bone defects: A systematic review. *Spine* 32, 824–30.

THE POWER OF ONE. (2014). Introduction. *Cell* 157, 3.

REDDI, A.H. (1994). Symbiosis of biotechnology and biomaterials: Applications in tissue engineering of bone and cartilage. *J Cell Biochem* 56(2), 192–95.

REDDI, A.H. (1997). Bone morphogenesis and modeling: Soluble signals sculpt osteosomes in the solid state. *Cell* 89, 159–61.

REDDI, A.H. (2000). Morphogenesis and tissue engineering of bone and cartilage: Inductive signals, stem cells, and biomimetic biomaterials. *Tissue Eng* 6(4), 351–59.

REDDI, A.H. (2005). BMPs: From bone morphogenetic to body morphogenetic proteins. *Cytokine Growth Factor Rev* 16(3), 249–50.

RIPAMONTI, U. (2004). Soluble, insoluble and geometric signals sculpt the architecture of mineralized tissues. *J Cell Mol Med* 8(2), 169–80.

RIPAMONTI, U. (2005). Bone induction by recombinant human osteogenic protein-1 (hOP-1, BMP-7) in the primate *Papio ursinus* with expression of mRNA of gene products of the TGF-$\beta$ superfamily. *J Cell Mol Med* 9, 911–28.

RIPAMONTI, U. (2006). Soluble osteogenic molecular signals and the induction of bone formation. *Biomaterials* 27, 807–22.

RIPAMONTI, U. (2009). Biomimetism, biomimetic matrices and the induction of bone formation. *J Cell Mol Med* 13(9B), 2953–72.

RIPAMONTI, U. (2012). BMPs/TGF-$\beta$ superfamily: Challenges for clinical translation. Presented at the 9th International Conference on Bone Morphogenetic Proteins, Lake Tahoe, CA.

RIPAMONTI, U., DIX-PEEK, T., PARAK, R., MILNER, B., DUARTE, R. (2015). Profiling bone morphogenetic proteins and transforming growth factor-$\beta$s by hTGF-$\beta_3$ pre-treated coral-derived macroporous constructs: The power of one. *Biomaterials*. doi: 10.1016/j.biomaterials.2015.01.058.

RIPAMONTI, U., DUARTE, R., FERRETTI, C. (2014a). Re-evaluating the induction of bone formation in primates. *Biomaterials* 35(35), 9407–22.

RIPAMONTI, U., DUNEAS, N., VAN DEN HEEVER, B., BOSCH, C., CROOKS, J. (1997). Recombinant transforming growth factor-$\beta_1$ induces endochondral bone in the baboon and synergizes with recombinant osteogenic protein-1 (bone morphogenetic protein-7) to initiate rapid bone formation. *J Bone Miner Res* 12, 1584–95.

RIPAMONTI, U., FERRETTI, C. (2002). Mandibular reconstruction using naturally-derived bone morphogenetic proteins: A clinical trial report. In T.S. Lindholm (ed.), *Advances in Skeletal Reconstruction Using Bone Morphogenetic Proteins*. Singapore: World Scientific, pp. 277–89.

RIPAMONTI, U., FERRETTI, C. (2010). The induction of bone formation by bone morphogenetic proteins in non-human and human primates. Presented at the 8th International Conference on Bone Morphogenetic Proteins, Leuven, Belgium.

RIPAMONTI, U., FERRETTI, C. (2012). Grand challenges for craniomandibulofacial reconstruction by human recombinant transforming growth factor-$\beta$3. In R.S. Tuan, F. Guilak, and A. Atala (eds.), *Keystone Symposia on Regenerative Tissue Engineering*, Breckenridge, CO. http://www.keystonesymposia.org.

RIPAMONTI, U., FERRETTI, C., HELIOTIS, M. (2006). Soluble and insoluble signals and the induction of bone formation: Molecular therapeutics recapitulating developnent. *J Anat* 209, 447–68.

RIPAMONTI, U., FERRETTI, C., HELIOTIS, M. (2008a). Soluble molecular signals and the induction of bone formation. In B. Guyuron, E. Eriksson, and J.A. Persing (eds.), *Plastic Surgery: Indication and Practice*. St. Louis, MO: Elsevier Science Global Medicine, pp. 363–74.

RIPAMONTI, U., FERRETTI, C., TEARE, J., BLANN, L. (2009). Transforming growth factor-$\beta$ isoforms and the induction of bone formation: Implications for reconstructive craniofacial surgery. *J Craniofac Surg* 20, 1544–55.

RIPAMONTI, U., HELIOTIS, M., FERRETTI, C. (2007). Bone morphogenetic proteins and the inducton of bone formation: From laboratory to patients. *Oral Maxillofac Surg Clin North Am* 19, 575–89, vii.

RIPAMONTI, U., KLAR, R.M. (2010). Regenerative frontiers in craniofacial reconstruction: Grand challenges and opportunities for the mammalian transforming growth factor-$\beta$ proteins. *Frontiers Physiol Craniofac Biol*. doi: 10.3389/fphys.2012.00172, 1, 1–7.

RIPAMONTI, U., KLAR, R.M., DUARTE, R., DIX-PEEK, T. (2014b). In primates, hTGF-$\beta_3$ initiates bone induction by up-regulating endogenous *BMPs* and is blocked by hNoggin. Presented at the 10th International Conference on Bone Morphogenetic Proteins, Berlin.

RIPAMONTI, U., KLAR, R.M., DUARTE, R., DIX-PEEK, T. (2014c). Engineering microenvironments superactivated by hTGF-$\beta_3$ reprogramming recruited differentiated pericytes into highly active secreting osteoblasts in primate striated muscles. Presented at the Proceedings of the Keystone Symposia on Engineering Cell Fate and Function: Stem Cells and Reprogramming, Olympic Valley, CA.

RIPAMONTI, U., KLAR, R.M., RENTON, L.F., FERRETTI, C. (2010). Synergistic induction of bone formation by hOP-1 and TGF-$\beta$3 in macroporous coral-derived hydroxyapatite constructs. *Biomaterials* 31(25), 6400–10.

RIPAMONTI, U., MA, S., CUNNINGHAM, N., YATES, L., REDDI, A.H. (1992). Initiation of bone regeneration in adult baboons by osteogenin, a bone morphogenetic protein. *Matrix* 12, 202–12.

RIPAMONTI, U., RAMOSHEBI L.N., PATTON J., MATSABA T., TEARE J., RENTON L. (2004). Soluble signals and insoluble substrata: Novel molecular cues instructing the induction of bone. In E.J. Massaro and J.M. Rogers (eds.), *The Skeleton*. Humana Press, Totowa, New Jersey, pp. 217–27.

RIPAMONTI, U., RAMOSHEBI, L.N., TEARE, J., RENTON, L., FERRETTI, C. (2008b). The induction of endochondral bone formation by transforming growth factor-$\beta_3$: Experimental studies in the non-human primate *Papio ursinus*. *J Cell Mol Med* 12(3), 1029–48.

Ripamonti, U., Teare, J., Ferretti, C. (2012). A macroscopic bioreactor super activated by the recombinant human transforming growth factor-$\beta_3$. *Frontiers Physiol* 3, 172. doi: 10.3389/fphys.2012.00172.

Ripamonti, U., van den Heever, B., Crooks, J., Tucker, M.M., Sampath, T.K., Rueger, D.C., Reddi, A.H. (2000). Long term evaluation of bone formation by osteogenic protein-1 in the baboon and relative efficacy of bone-derived bone morphogenetic proteins delivered by irradiated xenogeneic collagenous matrices. *J Bone Miner Res* 15, 1798–809.

Ripamonti, U., van den Heever, B., Sampath, T.K., Tucker, M.M. Rueger, D.C., Reddi, A.H. (1996). Complete regeneration of bone in the baboon by recombinant human osteogenic protein-1 (hOP-1, bone morphogenetic protein-7). *Growth Factors* 13, 273–89.

Sampath, T.K., Muthukumaran, N., Reddi, A.H. (1987). Isolation of osteogenin, an extracellular matrix-associated, bone-inductive protein, by heparin affinity chromatography. *Proc Natl Acad Sci USA* 84(20), 7109–13.

Sampath, T.K., Reddi, A.H. (1981). Dissociative extraction and reconstitution of extracellular matrix components involved in local bone differentiation. *Proc Natl Acad Sci USA* 78(12), 7599–603.

Sampath, T.K., Reddi, A.H. (1983). Homology of bone-inductive proteins from human, monkey, bovine, and rat extracellular matrix. *Proc Natl Acad Sci USA* 80(21), 6591–95.

Urist, M.R. (1965). Bone: Formation by autoinduction. *Science* 150(698), 893–99.

Williams, D.F. (2006). Tissue engineering: The multidisciplinary epitome of hope and despair. In R. Paton and L. McNamara (eds.), *Studies in Multidisciplinarity*. Amsterdam: Elsevier, pp. 483–524.

# CHAPTER SIX

# Synergistic Induction of Bone Formation by Relatively Low Doses of Transforming Growth Factor-$\beta_1$ and -$\beta_3$ in Binary Application with Recombinant Human Osteogenic Protein-1

*Ugo Ripamonti*

Bone Research Laboratory, School of Oral Health Sciences, Faculty of Health Sciences, University of the Witwatersrand, Johannesburg, Parktown, South Africa

## 6.1 Homologous but Molecularly Different Multiple Isoforms of the TGF-β Supergene Family Initiate the Induction of Bone Formation in Primate Species

Systematic studies by the Bone Research Laboratory of the University of the Witwatersrand, Johannesburg, on the bone induction principle (Urist et al. 1967, 1968) have yielded a series of published data on the induction of bone formation by naturally derived bone morphogenetic proteins (BMPs), the three mammalian transforming growth factor-β (TGF-β) isoforms, and calcium phosphate–based macroporous biomimetic matrices when implanted in heterotopic extraskeletal sites of the *rectus abdominis* muscle of adult Chacma baboons (*Papio ursinus*) (reviewed in Ripamonti 2004; Ripamonti et al. 2001, 2004, 2014). The systematic experimentation in *P. ursinus* has yielded unprecedented results, which were and still are against both the scientific and commercial dogmas of "*Bone: Formation by autoinduction*" (Urist 1965; Reddi and Huggins 1972).

In a number of systematic studies in different microenvironments of heterotopic intramuscular and orthotopic craniofacial sites, including the *rectus abdominis* muscle, the calvarium, and the mandible incorporating segmental full-thickness mandibular defects, as well as periodontally induced furcation defects, respectively, we have shown that primate tissues and microenvironments respond remarkably differently than rodents, lagomorphs, and canine tissues at identical doses of the soluble osteogenic molecular signals of the TGF-$\beta$ supergene family (Ripamonti 2003, 2006a, 2006b; Ripamonti et al. 2001, 2014).

As we describe at length in this volume, in the non-human primate *P. ursinus*, and possibly, by extension, to the primate *Homo sapiens*, the three mammalian TGF-$\beta$ isoforms induce rapid and substantial endochondral bone formation in heterotopic sites of the *rectus abdominis* muscle (Ripamonti et al. 1997, 2000, 2008; Ripamonti and Roden 2010a). This raises several questions with major significance to both our deeper understanding of the bone induction principle (Urist et al. 1967) and the optimal use of this knowledge for bone tissue engineering in clinical contexts (Ripamonti et al. 2009a). Why do the mammalian TGF-$\beta$ isoforms have such a different and critically important effect in non-human primates when compared to those of rodents or lagomorphs (Ripamonti 2006a; Ripamonti et al. 2009, 2014; Klar et al. 2014)? Which are the molecular and cellular differences that control endochondral bone formation by the three mammalian TGF-$\beta$ isoforms in *P. ursinus* versus rodents, lagomorphs, and canines? Chapters 3 and 4 describe our morphological and molecular studies, which have highlighted some of the mechanisms of the induction of bone formation by the hTGF-$\beta_3$ in the *rectus abdominis* muscle of *P. ursinus*, whereby selected *BMP* genes are expressed, and secreted gene products initiate the cascade of bone differentiation by induction (Ripamonti et al. 2014, 2015; Klar et al. 2014).

The scientific but particularly the commercial dogma is that single recombinant human BMPs do induce not only the induction of bone formation in the heterotopic bioassay in rodents, but also that a single recombinant morphogen is endowed with the unique capacity to regenerate full-thickness segmental defects of both the appendicular and craniofacial skeletons in preclinical and clinical contexts (reviewed by Ripamonti et al. 2006, 2007, 2014). Even the use of megadoses of recombinant human BMPs has, however, failed the induction of bone formation in clinical contexts (Ripamonti et al. 2007, 2014). Still, to date, the induction of bone formation in human patients has been dealt with in a rather crude, single-morphogen approach, and has resulted uninspiring clinical performance at massive (and expensive) doses (Ripamonti et al. 2007, 2014).

The Bone Research Laboratory and its scientific output has however never accepted this far too simplistic commercial vision of the bone induction principle but has rather envisioned since the purification of naturally-derived highly purified osteogenic fractions from baboon bone matrices (Ripamonti et al. 1992), the expression and secretion of several gene and gene products during the cascade of bone formation by induction (Ripamonti et al. 1992, Ripamonti et al. 2000b; Ripamonti et al. 2004; Ripamonti 2005). Genes and secreted gene products are then acting singly, synchronously, and synergistically to induce the complex molecular and cellular cascades of the induction of bone formation (Ripamonti et al. 2000b, 2001, 2004), culminating in the construction of heterotopic ossicles in extraskeletal intramuscular sites, the acid test to demonstrate "*Bone: Formation by autoinduction*" (Urist 1965; Reddi and Huggins 1972).

Uniquely in primates, the induction of bone formation initiates after the expression of *BMP* genes upon the heterotopic implantation of the three mammalian TGF-$\beta$ isoforms (Duneas et al. 1998; Ripamonti et al. 1997, 2000a, 2008, 2014, 2015; Klar et al. 2014). In the *bona fide* heterotopic assay for bone induction in rodents, the three mammalian TGF-$\beta$ isoforms, either purified from natural sources or expressed by recombinant techniques, do not initiate the heterotopic induction of bone formation (Ripamonti et al. 1997; Ripamonti 2003; Reddi 2000). The presence of several related, yet molecularly different, isoforms of the TGF-$\beta$ supergene family, with osteogenic activity in heterotopic *rectus abdominis* sites of *P. ursinus*, raises several important questions about the biological significance of this apparent redundancy of molecularly different, yet homologous, isoforms singly initiating the heterotopic induction of bone formation (Ripamonti et al. 1997).

The induction of bone formation in heterotopic extraskeletal sites of mammals by recombinant human decapentaplegic (DPP) and 60A of *Drosophila melanogaster* (Sampath et al. 1993), as reported in Chapter 4, has set back the molecular clock of the emergence of the vertebrates at almost 1 billion years before the present. The evolutionary conservation of the TGF-β supergene family members has evolved gene products of the fruit fly *D. melanogaster* to additionally initiate the induction of bone formation and skeletogenesis, the hallmark of the vertebrate mammals, thus predating the emergence of the bipedal hominids and the spectacular evolutionary speciation of *Homo* species across the planet (Ripamonti 2006a, 2009).

Singly, recombinant hBMPs initiate the cascade of bone differentiation by induction in the rodent heterotopic bioassay (Ripamonti 2005, 2006a; Ripamonti et al. 2005). The fact that single hBMPs initiate bone formation by induction does not preclude or indicate the requirements and interactions of other morphogens deployed singly, synchronously, and synergistically during the cascade of bone formation by induction, which then proceeds *via* the combined action of several BMPs resident within the natural *milieu* of the extracellular matrix (Ripamonti et al. 1997, 2000b, 2004; Ripamonti 2003, 2004). The presence of multiple molecular forms with osteogenic activity has also pointed to synergistic interactions during the induction of bone formation in both embryonic and postnatal development and morphogenesis (Ripamonti et al. 1997, 2004, 2008; Ripamonti 2003, 2006a).

## 6.2 Tissue Morphogenesis and Synergistic Interaction by Interposed Homologous Recombinant Morphogens

In a first set of experiments in four adult *P. ursinus*, lyophilized pellets of inactive insoluble collagenous bone matrix combined with doses of hOP-1 or hTGF-β₁ were implanted in an alternate fashion in a series of intramuscular pouches longitudinally prepared every 3 cm in the *rectus abdominis* muscle. During the course of the study, macroscopic and morphological analyses on day 30 showed the induction of large corticalized ossicles across the *rectus abdominis* muscle, transforming all treatment modalities, including hOP-1, hTGF-β₁, and control/treated collagenous bone matrix implants, into large corticalized ossicles expanding within the *rectus abdominis* muscle (Figure 6.1). The induction of corticalized ossicles (Figures 6.1 and 6.2) occasionally showed the developmental induction of cartilage differentiation mimicking the generation of embryonic growth plates (Figure 6.2a and b) (Ripamonti et al. 1997).

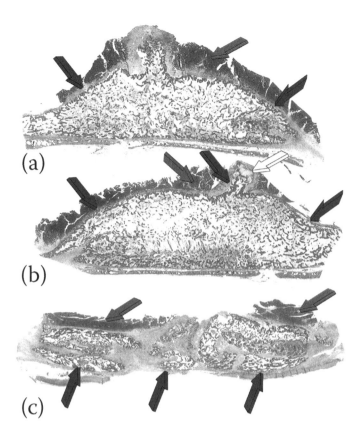

**FIGURE 6.1**  Tissue induction and morphogenesis by interposed recombinant morphogens across the *rectus abdominis* muscle. In a first set of experiments in four adult Chacma baboons (*Papio ursinus*), 100 mg of bovine insoluble collagenous bone matrix recombined with 5, 25, and 125 μg of osteogenic protein-1 (hOP-1) and 100 mg of baboon insoluble collagenous bone matrix recombined with recombinant human transforming growth factor-$\beta_1$ (hTGF-$\beta_1$) were implanted, separately, in the *rectus abdominis* muscle, and generated tissues, harvested on day 30, were subjected to histomorphometric analyses on undecalcified sections cut at 4 μm (Ripamonti et al. 1997). (a, b) Photomicrographs of large corticalized heterotopic ossicles (dark blue arrows) expanding within the *rectus abdominis* muscle (magenta arrows in a–c). White arrow in (b) indicates a large zone of chondrogenesis protruding within the *rectus abdominis* muscle. (c) Low power microphotograph of a large specimen block showing the generation of multiple corticalized ossicles across the *rectus abdominis* (dark blue arrows) and the induction of bone formation by interposed recombinant morphogens across the muscle (magenta arrows). Undecalcified sections embedded in K-Plast resin cut at 6 μm, stained and stained free floating. Original magnification: ×3.2 (a, b), ×2.2 (c).

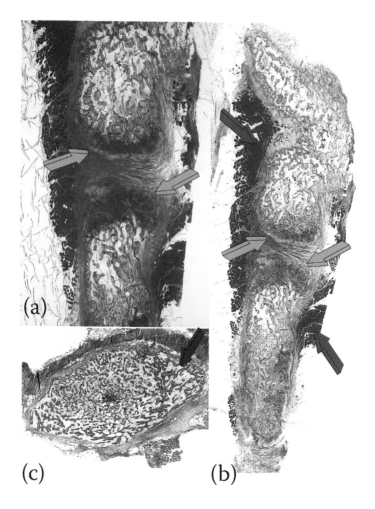

**FIGURE 6.2** Recapitulating development and morphogenesis by interposed recombinant morphogens across the *rectus abdominis* muscle of *Papio ursinus*. (a, b) Induction of large corticalized ossicles with the generation of cartilage zones (light blue arrows) at the site of two juxtaposed ossicles constructed within the *rectus abdominis* muscle (magenta arrows). There is a gradient of morphological structures highly suggestive of the induction of a rudimentary growth plate, with maturational gradients of tissue morphogenesis at the periphery of the newly formed mineralized ossicles, with induction of significant chondrogenesis surfacing trabeculations of the newly formed ossicles biomimetizing a cartilage growth plate (light blue arrows). (c) Induction of a large corticalized, mineralized (dark blue arrow) ossicle generated by 5 µg of recombinant human transforming growth factor-$\beta_1$ (hTGF-$\beta_1$) implanted in a different set of animals without hOP-1 to confirm the endochondral osteoinductivity of the recombinant hTGF-$\beta_1$ isoform (Ripamonti et al. 1997). Undecalcified sections embedded in K-Plast resin cut at 6 µm, stained and stained free floating. Original magnification: ×7.7 (a) ×17 (b) ×3.7 (c).

Morphological data, as shown in Figures 6.1 and 6.2, suggested a long-range activity of both morphogens caused by either a diffusion gradient or sequential inductive events between the sites of implanted morphogens, pointing to the synergistic interaction between hOP-1 and hTGF-$\beta_1$ in tissue induction and morphogenesis (Ripamonti et al. 1997).

A notable feature of the synergistic interaction by interposed morphogens (recombinant hOP-1 and hTGF-$\beta_1$) was the induction of large ossicles with zones of cartilage and peripheral corticalization, indicating that hTGF-$\beta_1$ and hOP-1 cooperate in the developmental processes responsible for the final sculpture of ossicles, complete with bone marrow permeating the newly formed trabeculae of bone (Figure 6.1a and b) (Ripamonti et al. 1997). The maturational gradient of tissue organization within generated ossicles has indicated rapid morphogenesis at the periphery of the ossicles, with corticalization of a mineralized shell of bone and with trabeculae of newly formed bone covered by thick osteoid seams populated by contiguous osteoblasts (Figure 6.1a and b).

The presence of voids and scattered residual matrix particles in the center of large ossicles has suggested constraints on ossicle growth as a result of limited central vascular invasion and cell ingrowth, as discussed in Chapter 3, highlighting the rapid induction of bone formation by the third mammalian TGF-$\beta$ isoform.

The morphogenesis of structurally organized cartilage zones, highly reminiscent of rudimentary embryonic growth plates (Figures 6.2a and b), is a finding that vividly illustrates the concept that regeneration of cartilage and bone in postnatal life shares common molecular and cellular mechanisms with embryonic bone development, and that the memory of developmental events in the embryo can be redeployed postnatally by the application of morphogen combinations (Ripamonti et al. 1997). We have previously indicated that the synergistic interaction among soluble morphogenetic signals could be a general principle adopted in embryonic development (Ripamonti et al. 1997). During embryogenesis, several molecularly different, yet homologous, soluble signals are deployed singly, synchronously, and synergistically to induce pattern formation and the establishment of tissue form and function or morphogenesis (Ripamonti et al. 1997; Ripamonti 2003, 2009), as exemplified by experiments showing a synergistic interaction of TGF-$\beta_1$ with basic fibroblast growth factor in the regulation of chondrogenesis during the murine otic capsule formation (Frenz et al. 1994).

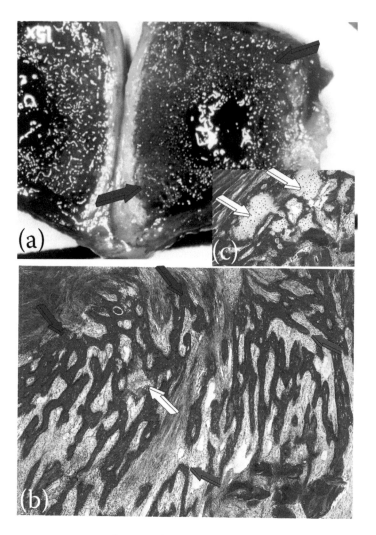

**FIGURE 6.3** Synergistic induction of tissue morphogenesis by binary application of recombinant human osteogenic protein-1 (hOP-1) with selected low doses of recombinant human transforming growth factor-$\beta_1$ (hTGF-$\beta_1$) (0.5 μg of hTGF-$\beta_1$). In a second set of experiments in an additional six adult Chacma baboons (*Papio ursinus*), and in an additional male baboon to solely test the inductive activity of hTGF-$\beta_1$, hOP-1, and hTGF-$\beta_1$, singly or in combination were added to 100 mg of bovine insoluble collagenous bone matrix and lyophilized to form solid pellets highly suitable for implantation in the *rectus abdominis* muscle. Pellets were inserted intramuscularly, resting over the dorsal fascia of the muscle, maximizing the distance between implants (Ripamonti et al. 1997). Generated tissue constructs were harvested on days 15 and 30 and processed for histological and histomorphometrical analyses (Ripamonti et al. 1997). (a) Induction of a large corticalized ossicle (dark blue arrows) by binary application of 25 μg of hOP-1 and 0.5 μg of hTGF-$\beta_1$ harvested on day 15 after implantation in intramuscular heterotopic sites with significant vascularization and bone marrow formation. (b) Multiple trabeculation of mineralized

The extensive induction of bone formation in the first set of experiments at sites of interposed morphogens (Figures 6.1 and 6.2) is possibly explained by desorption of the morphogens from the collagenous matrix carrier, as suggested by previous experiments in the primate (Ripamonti et al. 1996, 2000b), with subsequent diffusion of hOP-1 and hTGF-$\beta_1$ into the extracellular space away from the implanted collagenous matrix as carrier. Alternatively, the observed extended range of action is due to a diffusion gradient or the initiation of a sequential chain of cellular induction (Slack 1987; Michael Jones and Smith 1998; Lander 2007) that rapidly transfigures the *rectus abdominis* muscle into bone *in vivo*.

## 6.3 Synergistic Induction of Bone Formation

Experiments were thus performed with a modified implantation design aimed to maximize the interspecific distances between implants: in each animal, ipsilateral pouches, each separated by at least 8 and 4 cm of vertical and lateral intervening muscle, respectively (forming a zigzag implantation design), were implanted with specific combinations of hOP-1 and hTGF-$\beta_1$ in binary application of 25 µg of hOP-1 with 0.5 and 1.5 µg of hTGF-$\beta_1$; 5 µg of hTGF-$\beta_1$ was implanted singly at least 10 cm away from the lower doses of hOP-1 and in an additional *P. ursinus* animal without any hOP-1 implants (Figure 6.2c) (Ripamonti et al. 1997). Larger ossicles formed in heterotopic intramuscular sites with corticalization of the newly formed bone and bone marrow formation by day 15 after intramuscular implantation of binary application of 25 µg of hOP-1 with 0.5 and 1.5 µg of hTGF-$\beta_1$ (Figure 6.3a and b). On day 30, maximal response occurred at doses of 25 µg of hOP-1 in combination with 1.5 µg of hTGF-$\beta_1$ representing an optimal ratio of 20:1 by weight (Ripamonti et al. 1997, 2014a). Uniquely in the non-human primate *P. ursinus*, 5 µg of hTGF-$\beta_1$ delivered by insoluble collagenous bone matrix yielded large corticalized

---

**FIGURE 6.3** *Continued*

newly formed bone (dark blue arrows) covered by osteoid seams (magenta arrows) populated by contiguous osteoblasts with scattered chondrogenesis area (white arrow) within the trabeculae of newly formed mineralized bone. (c) Inset in (a) highlighting substantial chondrogenesis (white arrows) during endochondral bone differentiation by the synergistic induction of bone formation by binary application of 20:1 hOP-1/hTGF-$\beta_1$ (Ripamonti et al. 1997). Undecalcified sections embedded in K-Plast resin cut at 6 µm stained free-floating with Goldner's trichrome. Original magnification: ×27 (b), ×75 (c).

heterotopic ossicles on day 30 after heterotopic implantation in the *rectus abdominis* muscle (Figure 6.2c).

To explore the direct functional relationships between hTGF-$\beta_1$ and hOP-1 during the induction of bone formation in heterotopic *rectus abdominis* sites, binary applications of the recombinant morphogens were implanted in an additional four adult *P. ursinus* animals (Ripamonti et al. 1997); the combination of the homologous yet molecularly different, proteins resulted in the rapid generation of large corticalized ossicles complete with marrow formation by day 15 after *rectus abdominis* implantation (Figure 6.3) (Ripamonti et al. 1997; Duneas et al. 1998).

Evidence for synchronous developmental expression of *BMP*, *TGF*-$\beta$, and a variety of other TGF-$\beta$ family members' genes and gene products may help to design molecular therapeutic approaches based on recapitulation of embryonic development (Ripamonti et al. 1997); for one, the synergistic induction of bone formation engineers large ossicles in the non-human primate *P. ursinus*, requiring significantly fewer doses of singly applied recombinant hBMPs (in context the hOP-1 protein), with the overall induction of a greater amount of mineralized bone and osteoid synthesis in large corticalized ossicles *de novo* constructed in the *rectus abdominis* muscle of the Chacma baboon (*P. ursinus*) (Figure 6.4) (Ripamonti et al. 1997; Duneas et al. 1998).

Northern blot analyses on day 90 showed expression and autoinduction of OP-1 mRNAs after heterotopic implantation of the hOP-1 osteogenic devices (Figure 6.4c), together with expression of BMP-3 mRNA. Of note, Northern blot analyses showed a two- to threefold higher type IV collagen mRNA expression in synergistic constructs (Figure 6.4c, white arrow). Type IV collagen expression indicates the rapid sustained angiogenesis and capillary sprouting by the synergistically induced ossicles (Duneas et al. 1998; Ripamonti 2006a).

The rapid induction of bone formation by synergistic binary application of recombinant soluble signals with complete mineralization of the outer cortex with bone marrow formation bodes well for bone tissue engineering in the elderly, where repair phenomena are temporally delayed and tissue regeneration progresses slower than in younger patients.

Of note, binary application of hOP-1 with doses of porcine platelet-derived transforming growth factor-$\beta_1$ (pTGF-$\beta_1$), when implanted in orthotopic calvarial sites of *P. ursinus* (Figure 6.5), also yielded large corticalized ossicles by day 30, requiring far

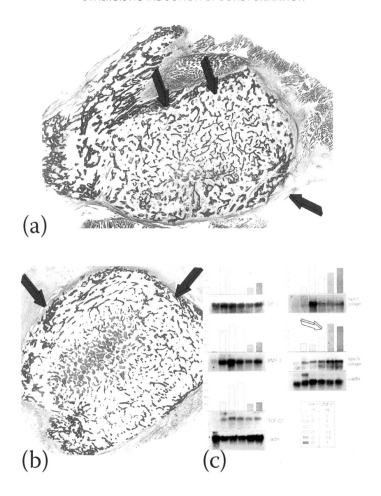

**FIGURE 6.4** Generated heterotopic ossicle by binary application of 25 µg of recombinant human osteogenic protein-1 (hOP-1) and 0.5 µg of recombinant human transforming growth factor-$\beta_1$ (hTGF-$\beta_1$) (Ripamonti et al. 1997). (a, b) Multiple trabeculation of mineralized newly formed bone with corticalization (dark blue arrows) of synergistically induced newly formed ossicles in heterotopic sites of the rectus abdominis muscle. Experiments were also run in additional *Papio ursinus* experimental animals to test the synergistic interaction of hOP-1 with porcine platelet-derived transforming growth factor-$\beta_1$ (pTGF-$\beta_1$). pTGF-$\beta_1$ was purified from 50 g of lyophilized platelets (Zymbio AS, Snåra, Norway) (Duneas et al. 1998) and tested singly or in binary application with 25 µg of hOP-1 (Duneas et al. 1998). (c) Northern blot analyses of hOP-1/pTGF-$\beta_1$ binary applications showed a two- to threefold increase of type IV collagen mRNA (white arrow in c), compared to morphogens implanted singly (Duneas et al. 1998); expression and synthesis of the type IV collagen gene product predate angiogenesis and capillary sprouting, tightly linked to osteogenesis in angiogenesis (Trueta 1963; Ripamonti 2006; Ripamonti et al. 2006).

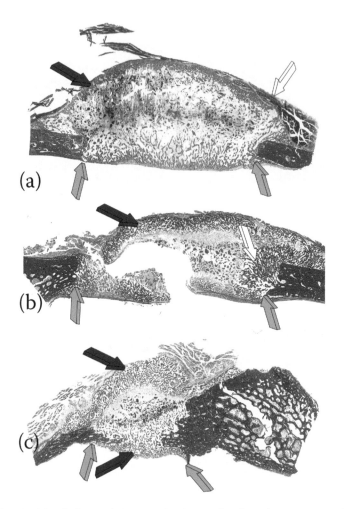

**FIGURE 6.5**  Morphology of tissue induction and calvarial regeneration by soluble morphogenetic signals, singly or in combination, implanted in non-healing calvarial defects of the Chacma baboon (*Papio ursinus*) recombined with insoluble collagenous bone matrices, harvested, and processed for undecalcified sectioning on day 30. (a) Extensive tissue induction by 0.5 mg of recombinant human osteogenic protein-1 (hOP-1) recombined with allogeneic insoluble collagenous bone matrix with dispersion of the collagenous matrix carrier within the newly formed tissue. Endocranial and pericranial osteogenetic fronts (dark blue arrow) with displacement of the temporalis muscle overlying the implanted defect. Note the scattered remnants of the collagenous matrix as carrier embedded within a loose, highly vascular cellular matrix that sustain significant osteogenesis at both the endocranial and pericranial fronts. (b, c) Morphology of tissue induction by the synergistic induction of bone formation by binary applications of 20 μg of hOP-1/5 μg of platelet-derived porcine transforming growth factor-$\beta_1$ (pTGF-$\beta_1$) (b) and 100 μg of hOP-1/15 μg of pTGF-$\beta_1$ (c). (b) Prominent substantial induction of bone formation by 25-fold less hOP-1 in binary application with 5 μg of pTGF-$\beta_1$. White arrow

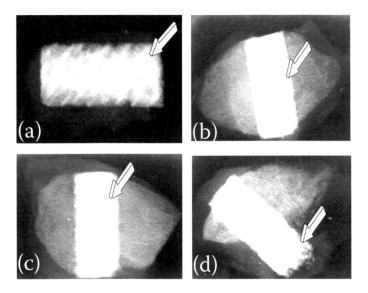

**FIGURE 6.6**   Digitized radiographic images of coral-derived macroporous constructs (white arrows) treated with binary application of 125 µg of osteogenic protein-1 (hOP-1) and 25 µg of recombinant human transforming growth factor-$\beta_3$ (hTGF-$\beta_3$) harvested on day 90 after heterotopic *rectus abdominis* implantation. (a) Macroporous bioreactor preloaded with 125 µg of hTGF-$\beta_3$ with no radiographic evidence of newly formed bone generating at the periphery of the macroporous construct. (b–d) Macroporous coral-derived bioreactors preloaded with binary applications of 125 µg of hOP-1 and 25 µg of hTGF-$\beta_3$. Substantial induction of mineralized trabeculated newly formed bone far exceeding the profile of the coral-derived bioreactors, the biological carrier used to deliver and initiate the synergistic induction of bone formation in intramuscular heterotopic sites (Ripamonti et al. 1997, 2010).

less quantities of hOP-1 to achieve equal regenerated constructs (Figure 6.5a and b). Calvarial ossicles displaced the *temporalis* muscle owing to the induction of prominent pericranial osteogenesis by day 30 after calvarial implantation; 20 µg of hOP-1 in binary application with 5 µg of pTGF-$\beta_1$ yielded corticalized, mineralized ossicles comparable to 0.5 mg of hOP-1 delivered by the identical collagenous bone matrix as carrier (Figure 6.5a

---

**FIGURE 6.5**   *Continued*

(b) indicates the substantial induction of bone formation flowing from a craniotomy edge invading the defect and expanding pericranially the induction of bone formation, thus displacing the temporalis muscle with the induction of prominent osteogenesis along the pericranial margin of the calvaria (dark blue arrow in b). (c) Significant osteogenesis by binary application of 100 µg of hOP-1/15 µg of pTGF-$\beta_1$ with the construction of a calvarial ossicle displacing both the pericranial and endocranial margins with induction of newly formed mineralized bone (dark blue arrows) with scattered remnants of the collagenous matrix as carrier.

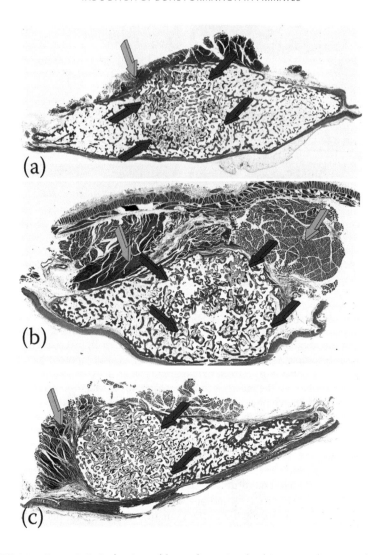

**FIGURE 6.7** Synergistic induction of bone formation by binary application of 125 μg of osteogenic protein-1 (hOP-1) and 25 μg of recombinant human transforming growth factor-β₃ (hTGF-β₃) preloaded onto coral-derived macroporous bioreactors and implanted in heterotopic sites of the *rectus abdominis* muscle of *Papio ursinus* (Ripamonti et al. 2010). (a–c) Massive induction of bone formation well outside the profile of the implanted macroporous bioreactors (dark blue arrows) extending within the *rectus abdominis* muscle (light blue arrows). The significant induction of bone formation at a distance from the implanted carrier for the recombinant morphogen is a classic morphological prerogative of the hTGF-β₃ isoform when implanted in heterotopic *rectus abdominis* sites of *P. ursinus*. As discussed in Chapter 3, systematic experimentation in *P. ursinus* with morphological analyses of undecalcified serial sections combined with molecular analyses of several bioreactors synergized by the recombinant hTGF-β₃ has indicated that the extended tissue induction and morphogenesis well outside the profile

and b). Of note, 100 μg of hOP-1 in binary application with 15 μg of pTGF-β₁ resulted in the induction of prominently mineralized endocranial and pericranial osteogenetic fronts displacing the temporalis muscle above the defects (Figure 6.5c).

Binary application of 125 μg of hOP-1 with 25 μg of hTGF-β₃ was preloaded onto coral-derived macroporous bioreactors and implanted in the rectus abdominis muscle of *P. ursinus* (Ripamonti et al. 2010). Generated tissue specimens were harvested 90 days after heterotopic intramuscular implantation. The *rectus abdominis* heterotopic microenvironment resulted in the construction of large mineralized ossicles far exceeding the dimensional profile of the implanted coral-derived macroporous constructs (Figure 6.6). Harvested newly formed heterotopic ossicles were radiographically digitized to show extensive induction of bone formation, dramatically exceeding the profile of the implanted macroporous constructs (Figure 6.6b–d). Because of the prominent synergistic induction of bone formation, specimen blocks were processed using the Exakt precision cutting and grinding system, and undecalcified sections cut at 30 μm were stained with Goldner's trichrome (Figure 6.7) (Ripamonti et al. 2010).

## 6.4 Temporal Expression of Markers of Bone Differentiation by Induction: Mechanistic Insights into the Synergistic Induction of Bone Formation

The relationship between BMPs and TGF-βs is indicative of the biological significance of redundancy during mammalian embryogenesis and postnatal tissue induction and morphogenesis (Thomadakis et al. 1999; Sampath et al. 1993; Ripamonti et al. 1997, 2006, 2014, 2015; Ripamonti 2007). The presence of several related, but molecularly different, isoforms endowed with osteogenic activity poses important questions about the biological significance of this apparent redundancy (Ripamonti 2004, 2006a; Ripamonti et al. 2008, 2014), indicating multiple interactions of several molecularly different, yet homologous,

---

**FIGURE 6.7** *Continued*

of the implanted heterotopic carriers is the result of the initiation of a sequential chain of cellular induction, possibly further potentiated by a diffusion gradient of the recombinant morphogen (Ripamonti et al. 1997, 2014, 2015). Undecalcified sections prepared by the Exakt diamond saw cutting and grinding system (Donath and Breuner 1982) ground and polished to 30 μm. Original magnification: ×2.7 (a–c).

isoforms deployed synchronously and synergistically during the cascade of bone formation by induction. The prominent induction of bone formation by binary application of a recombinant morphogenetic protein, in context, the recombinant hOP-1, with relatively low doses of recombinant hTGF-$\beta_1$ and -$\beta_3$, is mechanistically unresolved and needs to be assigned. How do relatively low doses of hTGF-$\beta_1$ and -$\beta_3$, when added to hOP-1 at the optimal ratio 1:20 by weight, profoundly set into motion the synergistic induction of bone formation?

The morphological analyses of several specimens of the synergistic induction of bone formation have shown constraints on ossicle growth as a result of limited central vascular and mesenchymal tissue invasion. As highlighted in Chapter 3, both TGF-$\beta_1$ and -$\beta_3$ act as chemotactic and mitogenic factors for responding precursor stem cells that are rapidly transformed into secreting osteoblastic cells when still at the periphery of the implanted specimens delivered by either insoluble collagenous bone matrices or macroporous coral-derived bioreactors (Ripamonti et al. 1997, 2008, 2014, 2015; Klar et al. 2014).

Of interest, it has been reported that exposure to hBMP-2 alters TGF-$\beta$ binding profiles, prompting bone cells toward the next step of phenotypic expression, simultaneously enhancing TGF-$\beta$-induced collagen synthesis and alkaline phosphatase activity, depending on the current state of bone cell differentiation and TGF-$\beta$ receptor expression, regulated by exposure to hBMP-2 (Centrella et al. 1995). The rapidity of tissue induction and transfiguration of the *rectus abdominis* muscle into bone *in vivo* (Chapter 3) is the most prominent prerogative of the synergistic induction of bone formation in the Chacma baboon (*P. ursinus*).

The recruitment of responding stem cells is clearly a phenomenon that results in the profound and rapid cascade of bone differentiation by induction when binary applications of recombinant morphogens are implanted in the *rectus abdominis* muscle of *P. ursinus*. The rapidity of cellular differentiation and tissue transfiguration *in vivo* is shown when doses of the hTGF-$\beta_3$ isoform are implanted in the *rectus abdominis* muscle of *P. ursinus*, combined with either insoluble collagenous bone matrix or coral-derived calcium phosphate–based bioreactors (Ripamonti et al. 2008, 2014, 2015; Klar et al. 2014).

The Bone Research Laboratory of the university has shown that hTGF-$\beta_3$ elicits the induction of bone formation by upregulating endogenous *BMP* expression (Ripamonti et al. 2014; Klar et al. 2014), as well the expression of other profiled *BMP* genes in further important experimentation in *P. ursinus* (Ripamonti et al. 2015). The induction of bone formation as initiated by

the hTGF-$\beta_3$ isoform is blocked by Noggin, a BMP antagonist (Ripamonti et al. 2014, 2015; Klar et al. 2014). We have
also shown a critical role for the hTGF-$\beta_3$ isoform in recruiting and reprogramming resident progenitor cells, including
pericytic/perivascular stem cells, into actively secreting osteoblastic cells (Ripamonti et al. 2014; Klar et al. 2014).

It is likely that *BMP* and *TGF*-$\beta$ genes expressed and profiled
upon hTGF-$\beta_3$ implantation result in the secretion of several *BMP*
and *TGF*-$\beta$ gene products that set into motion the rapid induction
of morphogenesis with muscle tissue transfiguration into bone
*in vivo*, the hallmark of the synergistic induction of bone formation (Ripamonti 2014; Ripamonti et al. 1997, 2014, 2015).

In previous studies using recombinant hOP-1, we have studied the temporal expression of markers of bone differentiation
by induction after heterotopic and orthotopic applications of
increasing doses of the hOP-1 osteogenic device (Ripamonti
2005). The temporal and spatial expression of TGF-$\beta_1$ mRNAs
with a relatively high expression on day 30, compared to low
expression patterns on days 15 and 90, has indicated the presence
of a specific temporal window during which the expression of
*TGF*-$\beta_1$ is mandatory for optimal osteogenesis (Ripamonti 2005).

Importantly, the manyfold increases in type IV collagen
mRNA synthesis (Figure 6.4c) over single applications of
hOP-1 and pTGF-$\beta_1$ have indicated that the two morphogens
interact synergistically to induce angiogenesis and vascular
invasion (Ripamonti et al. 1997; Duneas et al. 1998). Since
angiogenesis is a prerequisite for osteogenesis (Trueta 1963),
synergistically enhanced capillary sprouting and invasion,
including sinusoidal capillaries in the newly generated marrow tissue (Figure 6.3a), are part of the mechanisms whereby
hOP-1 and the mammalian TGF-$\beta_1$ and -$\beta_3$ isoforms synergize
to rapidly induce the induction of bone formation as early as
15 day postimplantation (Figure 6.3).

The induction of bone formation as initiated by the coral-
derived macroporous bioreactors has been a continuous source
of experimentation into the mechanisms of intramembranous
bone formation as initiated by the macroporous constructs
(Ripamonti 1991, 2009; Ripamonti et al. 1993, 2001, 2009b,
2010, 2012; Ripamonti and Roden 2010b; Klar et al. 2013).
Besides Noggin, previous experimentation in *P. ursinus* has
also shown that the induction of bone formation is additionally
abolished when coral-derived macroporous bioreactors are preloaded with doses of the osteoclastic inhibitor bisphosphonate
zoledronate Zometa® (Figure 6.8).

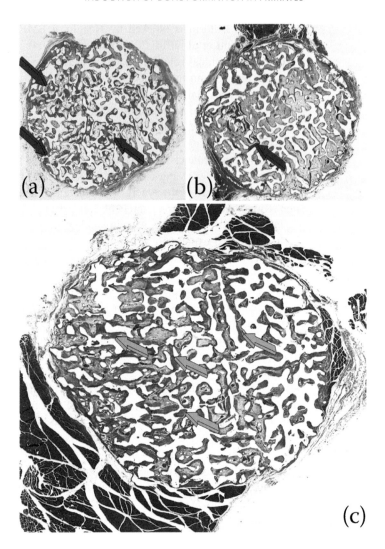

**FIGURE 6.8** Regulation of the induction of bone formation by coral-derived macroporous bioreactors by nanopatterned surface modifications by osteoclastogenesis (Ripamonti et al. 2010). (a) Induction of bone formation throughout the macroporous spaces (dark blue arrows) by 125 µg of human transforming growth factor-$\beta_3$ (hTGF-$\beta_3$) combined with coral-derived macroporous bioreactors implanted in the *rectus abdominis* muscle of *Papio ursinus*. (b) Untreated coral-derived macroporous construct induces the induction of bone differentiation across the macroporous spaces (dark blue arrow). (c) The spontaneous intrinsic induction of bone formation is abolished by preloading coral-derived macroporous constructs with 0.24 mg of bisphosphonate zoledronate Zometa (Ripamonti et al. 2010). Complete lack of bone differentiation with poorly constructed noninducing collagenous condensations (light blue arrows) (Ripamonti et al. 2010; Klar et al. 2013). Osteoclastogenesis and geometrically modified macroporous surfaces with Ca$^{++}$ released within the contained concavities of the macroporous

Zoledronate-treated macroporous constructs showed limited, if any, bone formation by induction (Figure 6.8c), and qRT-PCR showed prominent reduction of *OP-1* gene expression (Ripamonti et al. 2010). The lack of bone formation by zoledronate-treated specimens thus indicates that osteoclastogenesis is critical for the spontaneous induction of bone formation (Ripamonti et al. 2010). Micro- and macrotopographical surface modifications cut by osteoclasts after heterotopic intramuscular implantation of macroporous coral-derived constructs are regulators of the spontaneous and intrinsic induction of bone formation, as classically shown by the lack of bone formation in zoledronate-treated specimens (Figure 6.8c). Osteoclasts thus prepare functionalized surfaces for bone induction to occur, as during the remodeling cycle of the corticocancellous osteonic bone. Nanotopographic modifications imparted by the geometry of the functionalized surface thus control stem cell differentiation, osteogenic gene product expression, and the induction of bone formation (Ripamonti et al. 2010, 2012; Ripamonti 2012).

## 6.5 Synergistic Induction of Bone Formation in Full-Thickness Mandibular Defects of *P. ursinus*

The temporal window during which TGF-$\beta_1$ mRNA expression is mandatory for optimal osteogenesis (Ripamonti 2005) has been unambiguously demonstrated by the endochondral osteoinductivity of the three mammalian TGF-$\beta$ isoforms in *P. ursinus* (Ripamonti et al. 1997, 2000a, 2008; Ripamonti and Roden 2010a). The synergistic induction of bone formation (Ripamonti et al. 1997; Duneas et al. 1998) is Nature's strategy to rapidly and efficiently initiate tissue induction and morphogenesis, also providing a realistic therapeutic approach to tissue induction in clinical contexts (Ripamonti et al. 1997, 2007; Ripamonti 2006a; Ripamonti and Ferretti 2012).

Further studies aimed at potential therapeutic applications in clinical contexts showed that 125 μg of hTGF-$\beta_3$ per gram of

---

**FIGURE 6.8** *Continued*

spaces initiate mesenchymal stem cell differentiation, angiogenesis, osteoblastic-like cell differentiation with expression, synthesis, and secretion of osteogenic gene products, later embedded within the macroporous surfaces initiating the induction of bone formation as a secondary response (Klar et al. 2014; Ripamonti 2009; Ripamonti et al. 2009, 2010). Decalcified sections cut at 6 μm. Original magnification: ×3.7 (a–c).

insoluble collagenous bone matrix as carrier induces significant osteogenesis in full-thickness mandibular segmental defects in *P. ursinus* with unprecedented *restitutio ad integrum* of the buccal and lingual plates as early as 30 days after implantation (Ripamonti 2006b; Ripamonti et al. 2008).

Translational research from the benchtop to the clinical bedside using the hTGF-$\beta_3$ osteogenic is presented in Chapter 4. In contexts, binary applications of 125 µg of hTGF-$\beta_3$ with 2.5 mg of hOP-1 were also implanted in mandibular defects of *P. ursinus* (Vafaei 2014). Tissue specimens harvested 6 months after implantation showed the induction of remodeled solid blocks of osteonic bone regenerating the surgically created full-thickness segmental defects (Figure 6.9).

## Acknowledgments

The synergistic induction of bone formation by combining the mammalian transforming growth factor-$\beta_1$ (and -$\beta_3$) with recombinant human osteogenic protein-1 has been a fascinating discovery after the opening of the Bone Research Laboratory in January 1994 at the medical school of the University of the Witwatersrand, Johannesburg. The author of this CRC Press volume and chapter on the synergistic induction of bone forma-tion would like to thank the University of the Witwatersrand, Johannesburg, the National Research Foundation, and *ad hoc* grants of the Bone Research Laboratory for their support of the continuous and systematic studies in non-human primates since the early 1990s on the induction of bone formation by the mammalian transforming growth factor-$\beta$ isoforms. I thank the several students, laboratory technologists, scientists, and visit-ing professors who have significantly contributed to the sev-eral discoveries highlighted in this book, in particular Barbara van den Heever, Laura Yeates, Rooki Parak, Manolis Heliotis, Roland Klar, Carlo Ferretti, Raquel Duarte, Caroline Dickens, Therese Dix-Peek, and Brenda Milner. A special word of thanks to M.R. Urist, A.H. Reddi and "*Bone: Formation by autoinduction*" for continuously providing the inspiration to create, discover, and write.

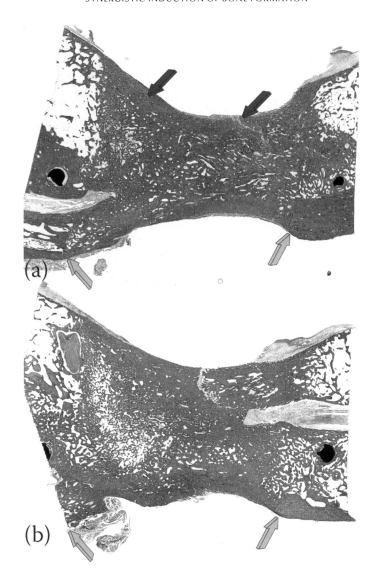

**FIGURE 6.9** *Restitutio ad integrum* of full-thickness segmental mandibular defects (light blue arrows) in *Papio ursinus* implanted with binary applications of 2.5 mg of recombinant human osteogenic protein-1 (hOP-1) with 125 µg of recombinant human transforming growth factor-$\beta_3$ (hTGF-$\beta_3$) harvested 6 months after implantation. Solid blocks of remodeled bone (dark blue arrows) regenerated the full-thickness mandibular defects. (a, b) Undecalcified sections prepared by the Exakt diamond saw cutting and grinding system (Donath and Breuner 1982) ground and polished to 30 µm. Original magnification: ×1.7 (a, b).

# References

CENTRELLA, M., CASINGHINO, S., KIM, J., PHAM, T., ROSEN, V., WOZNEY, J., McCARTHY, T.L. (1995). Independent changes in type I and type II receptors for transforming growth factor-β induced by bone morphogenetic protein-2 parallel expression of the osteoblast phenotype. *Mol Cell Biol* 15, 3273–81.

DONATH, K., BREUNER, G. A method for the study of undecalcified bones and teeth with attached soft tissues—the "Sage Schliff" (sawing and grinding) technique. *J Oral Pathol* 11, 318–26.

DUNEAS, N., CROOKS, J., RIPAMONTI, U. (1998). Transforming growth factor-β₁: Induction of bone morphogenetic proteins gene expression and synergistic interaction with osteogenic protein-1 (BMP-7). *Growth Factors* 15, 259–77.

FRENZ, D.A., LIU, W., WILLIAMS, J.D., HATCHER, V., GALINOVIC-SCHWARTZ, V., FLANDERS, K.C., VAN DE WATER, T.R. (1994). Induction of chondrogenesis: Requirement for synergitic intercation of basic fibroblast growth factor and transforming growth factor-beta. *Development* 120, 415–24.

KLAR, R.M., DUARTE, R., DIX-PEEK, T., DICKENS, C., FERRETTI, C., RIPAMONTI, U. (2013). Calcium ions and osteoclastogenesis initiate the induction of bone formation by coral-derived macroporous constructs. *J Cell Mol Med* 17(11), 1444–57.

KLAR, R.M., DUARTE, R., DIX-PEEK, T., RIPAMONTI, U. (2014). The induction of bone formation by the recombinant human transforming growth factor-β₃. *Biomaterials* 35(9), 2773–88.

LANDER, A.D. (2007). Morpheus unbound: Reimagining the morphogen gradient. *Cell* 128, 245–56.

MICHAEL JONES, C., SMITH J.C. (1998). Establishment of a BMP-4 morphogen gradient by long-range inhibition. *Dev Biol* 194, 12–17.

REDDI, A.H. (2000). Morphogenesis and tissue engineering of bone and cartilage: Inductive signals, stem cells, and biomimetic biomaterials. *Tissue Eng* 6(4), 351–59.

REDDI, A.H., HUGGINS, C.B. (1972). Biochemical sequences in the transformation of normal fibroblasts. *Proc Natl Acad Sci USA* 69, 1601–5.

RIPAMONTI, U. (1991). The morphogenesis of bone in replicas of porous hydroxyapatite obtained from conversion of calcium carbonate exoskeleton of coral. *J Bone Joint Surg Am* 73, 692–703.

RIPAMONTI, U. (2003). Osteogenic proteins of the transforming growth factor-β superfamily. In H.L. Henry and A.W. Norman (eds.), *Encyclopedia of Hormones*. Academic Press, San Diego, CA, pp. 80–86.

RIPAMONTI, U. (2004). Soluble, insoluble and geometric signals sculpt the architecture of mineralized tissues. *J Cell Mol Med* 8(2), 169–80.

RIPAMONTI, U. (2005). Bone induction by recombinant human osteogenic protein-1 (hOP-1, BMP-7) in the primate *Papio ursinus* with expression of mRNA of gene products of the TGF-β superfamily. *J Cell Mol Med* 9, 911–28.

RIPAMONTI, U. (2006a). Soluble osteogenic molecular signals and the induction of bone formation. *Biomaterials* 27, 807–22.

RIPAMONTI, U. (2006b). The Marshall Urist Awarded Lecture. Bone: Formation by autoinduction. In S. Vukicevic and A.H. Reddi (eds.), *Proceedings of the 6th International Conference on Bone Morphogenetic Proteins*, Dubrovnik, Croatia.

RIPAMONTI, U. (2007). Recapitulating development: A template for periodontal tissue engineering. *Tissue Eng* 13, 51–71.

RIPAMONTI, U. (2009). Biomimetism, biomimetic matrices and the induction of bone formation. *J Cell Mol Med* 13(9B), 2953–72.

RIPAMONTI, U. (2012). The concavity: The "shape of life" and the control of cell differentiation. *Sci South Africa*. http://www.scienceinafrica.co.za/2012/ripamonti_bone.htm.

RIPAMONTI, U. (2014). Transfiguration of neoplastic tumoral masses into bone for superior surgical debridement. NRF Blue Skies Funding Instrument—Concept Notes, NRF Grant 93117.

RIPAMONTI, U., CROOKS, J., KHOALI, L., RODEN, L. (2009b). The induction of bone formation by coral-derived calcium carbonate/hydroxyapatite constructs. *Biomaterials* 30, 1428–39.

RIPAMONTI, U., CROOKS, J., MATSABA, T., TASKER, J. (2000a). Induction of endochondral bone formation by recombinant human transforming growth factor-$\beta_2$ in the baboon (*Papio ursinus*). *Growth Factors* 17(4), 269–85.

RIPAMONTI, U., DIX-PEEK, T., PARAK, R., MILNER, B., DUARTE, R. (2015). Profiling bone morphogenetic proteins and transforming growth factor-βs by hTGF-$\beta_3$ pre-treated coral-derived macroporous constructs: The power of one. *Biomaterials* 49, 90–102.

RIPAMONTI, U., DUARTE, R., FERRETTI, C. (2014). Re-evaluating the induction of bone formation in primates. *Biomaterials* 35(35), 9407–22.

RIPAMONTI, U., DUNEAS, N., VAN DEN HEEVER, B., BOSCH, C., AND CROOKS, J. (1997). Recombinant transforming growth factor-$\beta_1$ induces endochondral bone in the baboon and synergizes with recombinant osteogenic protein-1 (bone morphogenetic protein-7) to initiate rapid bone formation. *J Bone Miner Res* 12, 1584–595.

RIPAMONTI, U., FERRETTI, C. (2012). Grand challenges for craniomandibulofacial reconstruction by human recombinant transforming growth factor-$\beta_3$. In R.S. Tuan, F. Guilak, and A. Atala (eds.), *Keystone Symposia on Regenerative Tissue Engineering*, Brekenridge, CO. http://www.keystonesymposia.org.

RIPAMONTI, U., FERRETTI, C., HELIOTIS, M. (2006). Soluble and insoluble signals and the induction of bone formation: Molecular therapeutics recapitulating developnent. *J Anat* 209, 447–68.

Ripamonti, U., Ferretti, C., Teare, J., Blann, L. (2009a). Transforming growth factor-β isoforms and the induction of bone formation: Implications for reconstructive craniofacial surgery. *J Craniofac Surg* 20, 1544–55.

Ripamonti, U., Heliotis, M., Ferretti, C. (2007). Bone morphogenetic proteins and the induction of bone formation: From laboratory to patients. *Oral Maxillofacial Surg Clin N Am* 19, 575–89.

Ripamonti, U., Herbst N.-N., Ramoshebi, L.N. (2005). Bone morphogenetic proteins in craniofacial and periodontal tissue engineering: Experimental studies in the non-human primate *Papio ursinus*. *Cytokine Growth Factor Rev* 16, 357–68.

Ripamonti, U., Klar, R.M., Renton, L.F., Ferretti, C. (2010). Synergistic induction of bone formation by hOP-1 and TGF-β$_3$ in macroporous coral-derived hydroxyapatite constructs. *Biomaterials* 31(25), 6400–10.

Ripamonti, U., Ma, S., Cunningham, N., Yates, L., Reddi, A.H. (1992). Initiation of bone regeneration in adult baboons by osteogenin, a bone morphogenetic protein. *Matrix* 12, 202–12.

Ripamonti, U., Ramoshebi, L.N., Matsaba, T., Tasker, J., Crooks, J., Teare, J. (2001). Bone induction by bmps/ops and related family members in primates. *J Bone Joint Surg Am* 83A(Suppl 1, Pt 2), S116–27.

Ripamonti, U., Ramoshebi L.N., Patton J., Matsaba T., Teare J., Renton L. (2004). Soluble signals and insoluble substrata: Novel molecular cues instructing the induction of bone. In E.J. Massaro and J.M. Rogers (eds.), *The Skeleton*. Humana Press, Totowa, New Jersey, pp. 217–27.

Ripamonti, U., Ramoshebi, L.N., Teare, J., Renton, L., Ferretti, C. (2008). The induction of endochondral bone formation by transforming growth factor-β$_3$: Experimental studies in the non-human primate *Papio ursinus*. *J Cell Mol Med* 12(3), 1029–48.

Ripamonti, U., Roden, L. (2010a). Induction of bone formation by transforming growth factor-β$_2$ in the non-human primate *Papio ursinus* and its modulation by skeletal muscle responding stem cells. *Cell Prolif* 43, 207–18.

Ripamonti, U., Roden, L. (2010b). Biomimetics for the induction of bone formation. *Expert Rev Med Devices* 74(4), 469–79.

Ripamonti, U., Teare, J., Ferretti, C. (2012). A macroporous bioreactor superactivated by the recombinant human transforming growth factor-β3. *Front Physiol* 3, 172. doi: 10.3389/fphys.2012.00172.

Ripamonti, U., van den Heever, B., Crooks, J., Tucker, M.M., Sampath, K.T., Rueger, D.C., Reddi, A.H. (2000b). Long-term evaluation of bone formation by osteogenic protein-1 in the baboon and relative efficacy of bone-derived bone morphogenetic proteins delivered by irradiated xenogeneic collagenous matrices. *J Bone Miner Res* 9, 1798–809.

RIPAMONTI, U., VAN DEN HEEVER, B., SAMPATH, K.T., TUCKER, M.M., RUEGER, D.C. (1996). Complete regeneration of bone in the baboon by recombinant osteogenic protein-1 (hOP-1, bone morphogenetic protein-7). *Growth Factors* 13, 273–89.

RIPAMONTI, U., VAN DEN HEEVER, B., VAN WYK, J. (1993). Expression of the osteogenic phenotype in porous hydroxyapatite implanted extraskeletally in baboons. *Matrix* 13, 491–502.

SAMPATH, T.K., RASHKA, K.E., DOCTOR, J.S., TUCKER, R.F., HOFFMANN, F.M. (1993). *Drosophila* TGF-β superfamily proteins induce endochondral bone formation in mammals. *Proc Natl Acad Sci USA* 90, 6004–8.

SLACK, J.M.W. (1987). Morphogenetic gradients: Past and present. *Trends Biochem Sci* 12, 200–6.

THOMADAKIS, G., CROOKS, J., RUEGER, D., RIPAMONTI, U. (1999). Immunolocalization of bone morphogenetic protein-2, -3 and osteogenic protein-1 during murine tooth morphogenesis and other craniofacial structures. *European Journal of Oral Sciences* 107, 368–77.

TRUETA, J. (1963). The role of the vessels in osteogenesis. *J Bone Joint Surg Am* 45B, 402–18.

URIST, M.R. (1965). Bone: Formation by autoinduction. *Science* 150, 893–99.

URIST, M.R., DOWELL, T.A., HAY, P.H., STRATES, B.S. (1968). Inductive substrate for bone formation. *Clin Orthop Relat Res* 59, 243–83.

URIST, M.R., SILVERMAN, B.F., BURING, K., DUBUC, F.L., ROSEMBERG, J.M. (1967). The bone induction principle. *Clin Orthop Relat Res* 53, 59–96.

VAFAEI, N. (2014). Recombinant human transforming growth factor-β3 and bone morphogenetic protein-7 for the regeneration of segmental mandibular defects in *Papio ursinus*. Master of dentistry (MDent), Maxillofacial and Oral Surgery, Bone Research Laboratory.

# CHAPTER SEVEN

# Induction of Periodontal Tissue Regeneration by Recombinant Human Transforming Growth Factor-β₃ with and without Myoblastic/Pericytic Stem Cells in *Papio ursinus*

*Jean-Claude Petit and Ugo Ripamonti*

Bone Research Laboratory, School of Oral Health Sciences, Faculty of Health Sciences, University of the Witwatersrand, Johannesburg, Parktown, South Africa

## 7.1 Introduction

The induction of cementogenesis with *de novo* generation of Sharpey's fibers inserting into newly formed cementum and beyond, into highly mineralized, almost impermeable dentine is a problem central to periodontal tissue regeneration. The molecular bases of such regeneration are the soluble molecular signals of the osteogenic proteins of the transforming growth factor-β (TGF-β) supergene family. In addition to the induction of bone formation *per se*, the osteogenic proteins of the TGF-β supergene family act as soluble molecular signals for the induction of periodontal tissue regeneration, sculpting the multicellular mineralized structures of the periodontal tissues with functionally oriented periodontal ligament fibers into newly formed cementum (Ripamonti 2007). Notably, 75 μg of hTGF-$\beta_3$ in a growth factor–reduced Matrigel® matrix induces prominent cementogenesis, with cementoid matrix covered by cementoblasts when implanted in Class II furcation defects surgically constructed in mandibular molars of the non-human primate Chacma baboon (*Papio ursinus*). The newly formed periodontal ligament space is characterized by running fibers tightly attached to the cementoid surface, penetrating as mineralized fibers the newly formed cementum and inserting into dentine. Prominent angiogenesis characterizes the newly formed periodontal ligament space as induced by the hTGF-$\beta_3$ isoform when implanted in Class II furcation defects of *P. ursinus*. Newly sprouting capillaries are lined by cellular elements with condensed chromatin, indicating the rapid and sustained induction of angiogenesis. Importantly, for tissue engineering in clinical contexts, the inductive activity of hTGF-$\beta_3$ in Matrigel matrix is enhanced by the addition of autogenous morcellated fragments of *rectus abdominis* muscle, thus providing myoblastic, pericytic/perivascular stem cells for further tissue induction and morphogenesis by the hTGF-$\beta_3$ isoform. In primates and in primates only, the mammalian hTGF-$\beta$ isoforms are powerful inducers of endochondral bone formation. Doses of the hTGF-$\beta_3$ osteogenic device when combined with either allogeneic insoluble collagenous bone matrix or growth factor–reduced Matrigel matrix as carrier induce regeneration of Class II and III furcation defects in *P. ursinus* (Ripamonti et al. 2009a; Ripamonti et al. 2009b; Ripamonti and Petit 2009).

The induction of the complex tissue morphologies of the periodontal tissues develops as a mosaic structure in which the osteogenic proteins of the TGF-$\beta$ supergene family singly, synergistically, and synchronously initiate and maintain tissue induction and morphogenesis (Thomadakis et al. 1999; Ripamonti 2007). In primate species, the presence of several, but molecularly different, molecular isoforms with osteogenic activity highlights the biological significance of this apparent redundancy, and indicates multiple interactions during both embryonic development and bone regeneration in postnatal life. Relatively low doses of hTGF-$\beta_1$ and -$\beta_3$ (0.5, 1.5, and 2.5 μg) synergize with osteogenic protein-1 (hOP-1; also known as BMP-7) to induce massive ossicles in the *rectus abdominis* of *P. ursinus* (25 μg of hOP-1 in the optimal ratio of 20:1 hOP-1:hTGF-$\beta_1$). In context, binary applications of hOP-1 and hTGF-$\beta_3$ in growth factor–reduced Matrigel matrix induce rapid and substantial periodontal tissue regeneration. Such regeneration is, however, tempered by the anatomy of the furcation defect model, which does not allow for rapid growth, proliferation, vascular invasion, and expansion of newly formed tissue constructs by the synergistic induction of bone formation, particularly when additionally potentiated by the addition of myoblastic, pericytic/perivascular stem cells from morcellated fragments of the autogenous *rectus abdominis* muscle.

## 7.2 Origins

The most striking characteristic of Nature is the deployment of never-ending evolutionary molecular pathways, leading to novel tissue morphologies sustained by gene expression arrays and activation responsible for the genesis of highly specialized morphogenetic structures. Nature relies on common, yet limited, molecular mechanisms tailored to provide the emergence of specialized tissues and organs (Reddi 2000; Ripamonti 2003; Ripamonti et al. 2004). The mechanisms of postnatal tissue regeneration are surprisingly simple: first, tissue regeneration recapitulates events that occur in the normal course of embryonic development and morphogenesis, and second, both embryonic development and tissue regeneration are equally regulated by a select few highly conserved families of morphogenetic gene products or morphogens (Reddi 2000; Ripamonti 2003; Ripamonti et al. 2004), first defined by Turing as "form generating substances" (Turing 1952).

Several studies have highlighted that odontogenesis, or tooth development, is a complex morphogenetic process tightly controlled by a select few conserved molecular circuits (Chinsembu 2012). Master molecules of odontogenesis mostly belong to the transforming growth factor-β (TGF-β) supergene family, the bone morphogenetic/osteogenic proteins (BMPs/OPs), and the three mammalian transforming growth factor-β (TGF-βs) isoforms. Master molecules are also the fibroblast growth factor (FGF) isoforms, the transcription factors wingless integrated (Wnt) and sonic hedgehog (Shh). All of the above genes and secreted gene products play a crucial role in tooth initiation, morphogenesis, and differentiation (Thesleff 2006; Foppiano et al. 2007; Chinsembu 2012).

Millions of years of evolution and selected speciation at the Pleio–Pleistocene boundary have set the emergence of the extant *Homo sapiens* craniomandibulofacial complex. The evolution of dentition with its supportive periodontal apparatus has been one of Nature's complex, yet highly successful, evolutionary challenges to provide proper functional and mechanical

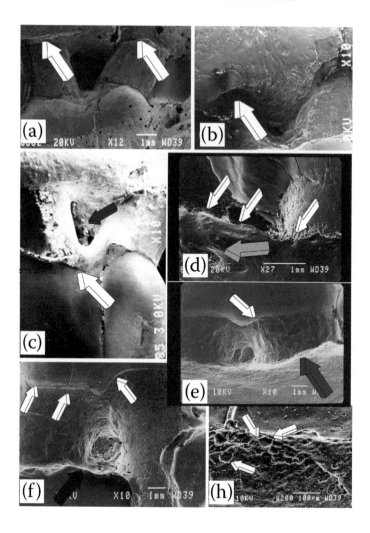

**FIGURE 7.1** Antiquity and severity of alveolar bone loss in fossilized gnathic remains of early hominids 2–3 million years before the present unearthed at Sterkfontein and Swartkrans, Blaauwbank Valley, South Africa. (a, b) Scanning electron microscopy (SEM) macrophotographs of dm1 and dm2 deciduous molars in a juvenile *Australopithecus africanus* specimen STS 24a unearthed at Sterkfontein, South Africa. (a) dm2 and (b) dm1 showing the hard evidence of severe alveolar bone loss exposing the roots of the affected deciduous molars by a suggested case of prepubertal periodontitis (Ripamonti 1988). White arrows indicate the remaining alveolar bony housings (a) with vertical cuneiform bone loss in the distal root of dm1 (white arrow in b). (c) Furcation exposure of dm2 with pronounced alveolar bone loss from the cementoenamel junction (white arrow) to the remaining alveolar bony housing with exposure of the furcation (dark gray arrow). Corticalization of the remaining interradicular bone with perforating Volkmann's canals (dark gray arrow). (d) SEM photograph of a mandibular specimen of adult *A. africanus* from Sterkfontein (STS 52) showing alveolar bone loss with exposure of the buccal

support during mastication, deglutition, and copulation, as well as many other physiological functions of human life (Reddi 1997), not least the extraordinary power of the human smile (Ripamonti 2009).

At the Pleio–Pleistocene boundary during hominid evolution and speciation, Nature evolved a more pathogenetic periodontal microflora during hominid phylogeny and speciation, together with subtle immunological differences after the emergence of the Homo clade (Ripamonti 1989). This eventually resulted in periodontal attachment loss in *Homo* species, following acute and chronic episodes of periodontitis (Ripamonti 1989). The observation of alveolar bone loss in Pliocene fossilized *Australopithecus africanus* gnathic remains provides paleopathological evidence of the antiquity of periodontal disease (Ripamonti 1989). By including the reported case of a prepubertal periodontitis in a juvenile *A. africanus* specimen (Ripamonti 1988), periodontal pathology is the only available evidence of a disease entity in Pliocene hominid remains (Figure 7.1) (Ripamonti 1988).

Attachment loss with associated bone loss with the inferred induction of acute and chronic periodontitis is thus the first recognized disease in hominid evolution (Ripamonti 1988). There is, however, a transition of morphological events from the available Australopithecinae fossilized material to the emerging *Homo* species, such as *Homo habilis* and *Homo erectus* unearthed at the Cradle of Mankind, Blaauwbank Valley, South Africa; *Homo* species show the fossilized evidence of alveolar bone loss with the emergence of a vertical component of attachment loss with crateriform osseous lesions, as found in extant *H. sapiens*, but not in *A. africanus* and *Australopithecus robustus* gnathic remains (Ripamonti 1989).

Speciation 2–4 million years before the present in a harsh and hostile environment, remarkably also set the induction of periodontitis in early hominids, primitive may be, yet

---

**FIGURE 7.1** *Continued*

furcation of the first mandibular molar with lipping of the residual bony housing (white arrows) with a perforating vascular canal (white arrow). (e) Adult *A. africanus* gnathic remains with vertical bone loss from the cementoenamel junction (white arrow) to the remaining buccal alveolar bony housing (dark gray arrow). (f) Complete exposure of the furca with clinically significant horizontal bone loss from the cementoenamel junction (white arrows) to the remaining alveolar bony housing (dark gray arrow) in adult *A. africanus* mandibular gnathic remains. (h) Higher magnification of the buccal root shown in (f) reveals a surface topography highly reminiscent of the polygonal pattern of insertion of Sharpey's fibers into cementum (white arrows) (Ripamonti et al. 1989).

irresistibly walking into evolutionary pathways of creativity, toward the spectacular growth of the cerebral hemispheres we have thus inherited (Ripamonti 2007).

Several million years after the Australopithecinae and *Homo* species roamed the Blaauwbank Valley in the great mother Africa, from whose wombs the early hominids originated, extant Homo identified novel soluble pleiotropic molecular signals to engineer periodontal tissue regeneration (Ripamonti 2007). The complex tissue morphologies of the periodontal tissues are a superior example of Nature's design and architecture in which the *continuum* between the soluble and insoluble signals of the extracellular matrix is regulated by signals in solution (Reddi 1997), interacting with extracellular matrices and responding stem cells of the periodontal tissues identified in paravascular/perivascular stem cell "niches" within the periodontal ligament space facing the alveolar bone and the cementum (Bartold et al. 2000; Lin et al. 2008; Ripamonti 2007; Ripamonti and Petit 2009).

The induction of bone formation requires three key components (Reddi 1997, 2000; Ripamonti et al. 2000a, 2004): soluble osteogenic molecular signals, responding stem cells, and insoluble signals or substrata. The latter is the scaffold upon which differentiating induced mesenchymal stem cells' erect "*Bone: Formation by autoinduction*" (Urist 1965). Last century's research has taught us a fundamental truth in bone tissue engineering and regenerative medicine at large: insoluble signals or substrata, when recombined or reconstituted with soluble osteogenic molecular signals, trigger the ripple-like cascade of tissue induction and morphogenesis (Sampath and Reddi 1981, 1983; Ripamonti and Reddi 1995; Reddi 2000; Ripamonti et al. 2000a, 2004).

The induction of bone formation by recombining or reconstituting soluble osteogenic molecular signals with insoluble signals or substrata has been pivotal for setting the rules of the regenerative medicine paradigm: the induction of tissue morphogenesis by combinatorial molecular protocols (Sampath and Reddi 1981, 1983; Khouri et al. 1991; Ripamonti and Reddi 1995; Reddi 2000; Ripamonti et al. 2004; Ripamonti et al. 2007) (Figure 7.2) whereby soluble and insoluble signals are combined to initiate the molecular and morphogenetic cascades of the induction of bone formation in heterotopic extraskeletal sites (Ripamonti et al. 2000a, 2004).

The induction of bone formation has evolved as the prototype of the tissue engineering paradigm, continuously developing the concept and further expanding the depths and breadth

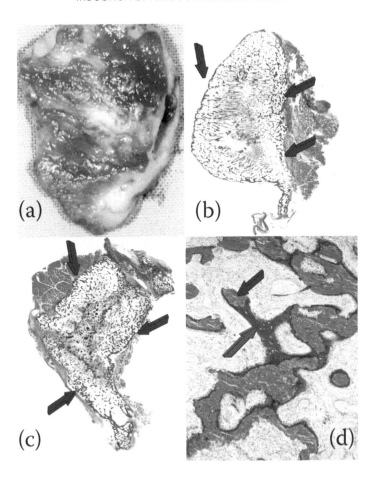

**FIGURE 7.2** *"Bone: Formation by autoinduction"* (Urist 1965). (a) *De novo* generation of large corticalized mineralized ossicles in the *rectus abdominis* muscle of adult Chacma baboons (*Papio ursinus*) 30 days after implantation of 125 µg of recombinant human transforming growth factor-$\beta_3$ (hTGF-$\beta_3$) recombined with allogeneic insoluble collagenous bone matrix. (b) Undecalcified section cut at 6 µm of the newly generated ossicle showing corticalization (dark blue arrows) of the newly formed bone. (c) Reproducible, newly formed ossicles with corticalization (dark blue arrows) in the *rectus abdominis* of *P. ursinus* using the 125 µg dose of the recombinant protein. (d) High-power view of undecalcified section cut at 6 µm as shown in (c), highlighting the induction of mineralized newly formed bone (dark blue arrow) surfaced by large osteoid seams (magenta arrow) throughout the specimen populated by contiguous osteoblasts facing a highly vascularized connective tissue matrix.

of novel developmental, biological, and surgical concepts. The induction of bone formation initiates by invocation of osteogenic soluble molecular signals that, when combined with insoluble signals or substrata, set into motion the induction of bone formation (Sampath and Reddi 1981; Reddi 2000; Ripamonti et al. 2000a, 2004; Ripamonti et al. 2007). The osteogenic soluble molecular signals of the TGF-β supergene family, the BMPs/OPs, and in primates only, the three mammalian TGF-β isoforms (Figure 7.2) induce endochondral bone formation as a recapitulation of embryonic development (Ripamonti and Reddi 1995; Ripamonti 2003; Ripamonti et al. 2004, 2008). Importantly, for tissue engineering in clinical contexts, neither the solubilized proteins nor the insoluble signal or residue is active (Sampath and Reddi 1981, 1983; Ripamonti and Reddi 1995; Reddi 2000; Ripamonti et al. 2004). The reconstitution or recombination of the soluble and the insoluble components of the chaotropically extracted bone matrix, however, restores the osteogenic activity when implanted in heterotopic subcutaneous sites of the rodent bioassay (Sampath and Reddi 1981, 1983; Ripamonti and Reddi 1995). This operational reconstitution of the soluble molecular signal with an insoluble signal or substratum was a key experiment that provided a bioassay for *bona fide* initiators of endochondral bone differentiation in extraskeletal sites of animal models (Reddi and Huggins 1972; Sampath and Reddi 1981, 1983; Khouri et al. 1991; Ripamonti and Reddi 1995; Reddi 2000).

## 7.3  Periodontal Tissue Regeneration by the Osteogenic Proteins of the TGF-β Supergene Family

The regenerative potential of bone, a highly vascular three-dimensional mineralized matrix that remodels throughout life and heals without scarring, has been shown since antiquity (Reddi 1981, 1984). Despite the regenerative capacity of bone, periodontal osseous defects, however, lack the template for orchestrated tissue regeneration. Periodontal tissue engineering has often proved to be elusive because of the challenges involved in tissue differentiation, migration, attachment, and spatial and temporal positioning of a variety of embryologically different cellular phenotypes on a supportive, yet completely avascular, mineralized dentinal substratum (Gottlow et al. 1984; Ripamonti and Reddi 1994, 1997; Ripamonti 2007).

Cellular activity is a vital component of comparative large bone defects' repair and regenerative processes. Striated muscle, containing several different stem cell niches (Zheng et al. 2007; Kovacic and Boehm 2009) harboring a variety of stem cells with osteogenic and myogenic phenotypic differentiating pathways, could be an adjuvant treatment to soluble molecular signals to induce regeneration in Class II and III furcation defects as a result of recurrent destructive episodes of advanced periodontitis.

The challenging problem of bone formation by induction in primates (Ripamonti 1991), and particularly the induction of cementogenesis with *de novo* generation of Sharpey's fibers inserted within the instrumented root surfaces (Ripamonti and Reddi 1994, 1997; Ripamonti 2007), has inspired the Bone Research Laboratory of the University of the Witwatersrand, Johannesburg, to create experimental animal models using the adult Chacma baboon (*P. ursinus*) that share osteonic bone remodeling close to that of man (Schnitzler et al. 1993). Preclinical data obtained in non-human primates are critical for translational research in clinical contexts, and several experimental studies in *P. ursinus* have shown that the osteogenic soluble molecular signals of the TGF-β supergene family are also the initiators of cementogenesis, inducing the assembly of a functionally oriented periodontal ligament system (Ripamonti et al. 2004, 2005; Ripamonti 2006, 2007). The osteogenic proteins of the TGF-β supergene family are indeed the soluble molecular signals that initiate the regeneration of the periodontal tissues, including the induction of cementogenesis (Ripamonti et al. 1994; Ripamonti and Reddi 1997; Ripamonti 2007; Ripamonti and Petit 2009).

To sculpt tissue morphogenesis, including the complex morphologies of the periodontal tissues that lock the teeth into the alveoli *via* the regeneration of cementum with the faithful insertion of Sharpey's fibers, Nature relies on common, yet limited, molecular mechanisms to sustain the emergence of specialized tissues and organs (Thesleff 2006; Foppiano et al. 2007; Ripamonti 2007; Chinsembu 2012). Tissue regeneration in postnatal life recapitulates events that occur in the normal course of embryonic development (Reddi 1981, 2000; Ripamonti 2007); both embryonic development and tissue regeneration are regulated by a select few and highly conserved families of morphogens (Thesleff 2006; Foppiano et al. 2007; Ripamonti 2007; Chinsembu 2012).

The above is shown by the mosaicism of expression and localization patterns of several osteogenic proteins of the TGF-β supergene family that regulate tooth morphogenesis at different stages of development as temporally and spatially connected events (Vainio et al. 1993; Hogan 1996; Thesleff and Nieminen 1996; Thesleff and Sharpe 1997; Åberg et al. 1997; Thomadakis et al. 1999).

*In situ* hybridization and immunolocalization of BMPs/OPs during developmental tooth morphogenesis indicate that the secreted gene products play morphogenetic roles during cementogenesis and the assembly of a functionally oriented periodontal ligament system (Vainio et al. 1993; Hogan 1996; Thesleff and Nieminen 1996; Thesleff and Sharpe 1997; Åberg et al. 1997; Thomadakis et al. 1999). BMPs/OPs regulate tooth morphogenesis at different stages of development as temporally and spatially connected events (Hogan 1996; Åberg et al. 1997; Ripamonti 2007). Nature's parsimony in sculpting tissue constructs is epitomized by the deployment of a restricted family of molecularly different but functionally homologous gene products with minor variations in amino acid sequence motifs within highly conserved C-terminal regions (Ripamonti and Reddi 1994; Reddi 2000; Ripamonti et al. 2004; Ripamonti 2007). The secreted proteins are endowed with the striking prerogative of initiating endochondral bone formation by induction, in addition to specialized pleiotropic functions controlled by selected amino acid motifs as set in the C-terminal domain (Thomadakis et al. 1999; Ripamonti 2007), which include the induction of cementogenesis with *de novo* generation of Sharpey's fibers.

Of note, the expression patterns of BMP-3, BMP-5, and BMP-6 are all completely different during tooth morphogenesis, with each BMP restricted to one specific cell lineage (Åberg et al. 1997). BMP-3 is intensely expressed in the mesial part of the dental papilla, and subsequently to different stages of differentiation, it became restricted to the dental follicle cells (Åberg et al. 1997). Importantly, these cells give rise to the periodontal tissues, including cementoblasts, which secret collagenic material, and thus the induction of cementogenesis (Thesleff and Nieminen 1996; Åberg et al. 1997; Thomadakis et al. 1999) along the forming dentine. Of interest, Åberg et al. (1997) indicated that BMP-3 might be involved in the formation of cementum since the reported data did not observe the expression of any other BMPs in the cementoblastic lineage (Åberg et al. 1997).

## 7.4 Periodontal Tissue Regeneration by Naturally Derived, Highly Purified and Recombinant Human Bone Morphogenetic Proteins

A set of experiments was initiated at the Bone Research Laboratory to test whether naturally derived, highly purified proteins extracted and purified from bovine and baboon bone matrices would initiate periodontal tissue induction and regeneration when implanted in Class II furcation defects of *P. ursinus* (Ripamonti et al. 1994). Our laboratories first tested highly purified osteogenic proteins purified greater than 50,000-fold from crude extracts of bovine bone matrices (Luyten et al. 1989). Osteogenic proteins, solubilized by chaotropic extraction of demineralized bone matrices with 6 M guanidinium hydrochloride (Gnd-HCl), were sequentially purified by hydroxyapatite adsorption and heparin–Sepharose affinity chromatography (Luyten et al. 1989; Ripamonti et al. 1992). The recovered 500 mM NaCl step-eluted fractions were exchanged and concentrated with 6 M Gdn-HCl and loaded onto tandem Sephacryl S-200 gel filtration chromatography columns (Luyten et al. 1989; Ripamonti et al. 1992). Eluted fractions with biological activity in the rodent subcutaneous assay were further concentrated and exchanged to 6 M urea for implantation in Class II furcation defects surgically prepared in adult Chacma baboons. Defects were implanted with 250 μg of highly purified, naturally derived bovine osteogenic proteins combined with allogeneic insoluble collagenous bone matrix as carrier (Ripamonti et al. 1992, 1994).

Morphological analyses on undecalcified sections cut at 3 μm showed the generation of functionally oriented Sharpey's fibers into dentine matrix, with condensed cellular elements initiating angiogenesis within the regenerated periodontal ligament space (Figure 7.3). Angiogenesis is a prominent morphological feature of periodontal tissue regeneration when using highly purified, naturally derived osteogenic proteins (Figure 7.3). High magnification of undecalcified sections shows periodontal ligament fibers merging with the basement membrane of invading sprouting capillaries. The newly formed capillaries are thus suspended within the regenerated periodontal ligament space (Figure 7.4). High-power views show Sharpey's fibers inserting as mineralized fibers directly into dentine, with mesenchymal cells "riding" individual periodontal fiber across morphogenetic gradients, that is, *versus* the alveolar bony side or the cementum side of the newly formed periodontal ligament space (Figures 7.3h and 7.4a) (Ripamonti et al. 2009a).

A high-power view of undecalcified sections shows the generation and insertion of Sharpey's fibers within the planed root surfaces covered by cementoblasts longitudinally aligned against the prepared root surfaces (Figure 7.3). Sharpey's fibers are seen inserting into the dentine matrix among cementoblasts

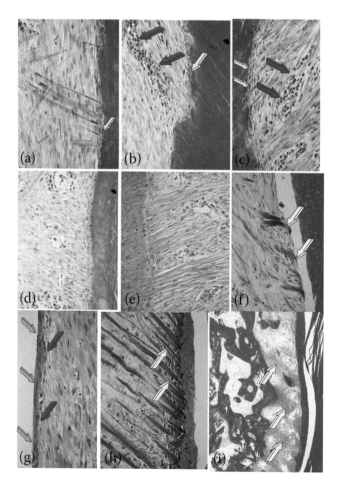

**FIGURE 7.3** Reconstructing cementum and periodontal ligament fibers inserted as Sharpey's fibers into newly formed cementum and dentine by the osteogenic proteins of the transforming growth factor-β (TGF-β) supergene family. Composite iconographic plate of periodontal tissue induction in mandibular furcation defects of the non-human primate *Papio ursinus*, digitizing tissue induction and morphogenesis as initiated by naturally derived, highly purified bone morphogenetic proteins (BMPs) purified greater than 50,000-fold from bovine bone matrices after hydroxyapatite Ultrogel adsorption, heparin–Sepharose affinity, and gel filtration chromatography onto tandem Sephacryl S-200, yielding highly purified osteogenic fractions with biological activity in the rodent subcutaneous bioassay (Luyten et al. 1989; Ripamonti et al. 1992). (a–c) *De novo* generation and faithful insertion of Sharpey's fibers into instrumented root dentine tightly attaching into the highly mineralized dentinal matrix, with cementoblasts aligned between single fibers secreting the very first cementoid matrix along the exposed root surfaces. Extensive cellular trafficking within the periodontal ligament space with cellular elements (magenta arrows in b and c), with condensed chromatin interpreted as angioblasts initiating angiogenesis and capillary sprouting within the newly formed ligament space. Note the vertical alignment of cementoblastic cells along the inserted fibers penetrating

and condensed cellular elements interpreted as angioblasts differentiating newly formed capillaries within the newly generated periodontal ligament space (Figure 7.3b–d). Capillary sprouting and invasion are prerequisite for osteogenesis (Trueta 1963) since both angiogenic and osteogenic proteins are bound to type IV collagen of the basement membrane of the invading capillaries (Vlodavsky et al. 1987; Folkman et al. 1988; Paralkar et al. 1990, 1991). The binding and sequestration of both angiogenic and osteogenic proteins to the basement membrane components of the invading capillaries provide the conceptual framework for the supramolecular assembly of the extracellular matrix of the periodontal ligament space. Angiogenic and bone morphogenetic proteins bound to type IV collagen of the basement membrane of the invading capillaries are presented in an immobilized form to responding mesenchymal cells to initiate osteogenesis in angiogenesis (Ripamonti 2006, 2007; Ripamonti et al. 2007).

The role of the vessels in osteogenesis has been lucidly defined by the classic morphological studies of Trueta (1963), who defined the invading sprouting capillaries as "osteogenetic vessels." Several osteogenetic vessels are formed in the developing periodontal ligament space, and the periodontal ligament fibers are intimately connected and inserted into the basement

---

FIGURE 7.3   *Continued*

within the mineralized dentine. The newly formed fibers directly inserting into dentine not only align, but also provide tridimensional support for stacking cementoblasts for continuous delivery to the exposed dentinal surface (white arrows in a–c), ultimately secreting cemental matrix. (d, e) High-power views of newly formed periodontal ligament space with pronounced angiogenesis and cellular trafficking as initiated by highly purified, naturally derived BMPs with Sharpey's fibers inserting directly into dentine. Note that the collagenic supramolecular assembly of the functional unit of Sharpey's fibers penetrates not only the newly formed cementoid and mineralized cementum, but also the highly mineralized avascular dentine matrix, thus providing a mechanically tight superior attachment bond to the dentine extracellular matrix substratum. (f) Generation of Sharpey's fibers and very early cementogenesis along the exposed root surfaces; nucleation of collagenic material into periodontal ligament fibers (white arrows) with early synthesis of a thin cementoid matrix beginning to mineralize (white arrows in f). (g–i) Cementogenesis, assembly of Sharpey's fibers, deposition of as yet to be mineralized cementoid matrix by secreting cementoblasts (magenta arrows in g). (h) Assembly of Sharpey's fibers along the newly synthesized mineralized cementum with progenitor stem cells riding single fibers (white arrows) from and to the cementoblastic or osteoblastic compartments of the newly formed periodontal ligament space. (i) Final assembly of periodontal ligament fibers connecting as bundles of highly specialized collagenic material (white arrows) the newly formed and mineralized cementum to the newly formed mineralized alveolar bone 60 days after implantation of highly purified, naturally derived osteogenic proteins (Ripamonti et al. 1994).

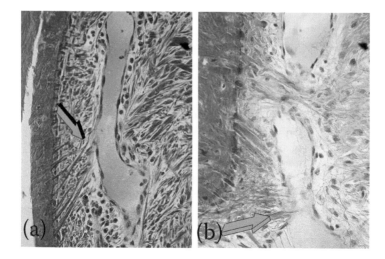

**FIGURE 7.4** Tissue induction and morphogenesis of the periodontal ligament space with osteogenetic vessels of Trueta's definition (Trueta 1963) that provide a continuous flow of progenitor stem cells from the vascular compartment to the cementogenic or osteogenic microenvironments of the newly formed periodontal ligament space. (a, b) Sharpey's fibers directly blending and inserting into the basement membrane structures of the vessels suspended within the periodontal ligament space (light blue arrows). (b) Fibers generating from the alveolar bone also insert into the capillaries (light blue arrow), thus providing a continuous flow of progenitor stem cells to both cementogenic and osteogenic compartments of the periodontal ligament space.

membrane of the osteogenetic vessels (Figure 7.4a and b), thus providing a three-dimensional supramolecular assembly for cellular trafficking from and to the endothelial/pericytic compartments and the alveolar bone and the cementum side of the periodontal ligament space (Ripamonti et al. 1994, 2009a; Ripamonti 2007).

High-power digital images of undecalcified sections cut at 3 μm of the periodontal ligament space of furcation defects treated with 250 μg of highly purified, naturally derived osteogenic fractions (Ripamonti et al. 1994) show that periodontal ligament and Sharpey's fibers provide the supramolecular assembly for progenitor pericytic cells to ride the fibers from and to the osteogenetic vessels (Figures 7.3h and 7.4). The osteogenetic vessels of Trueta's definition (1963) are thus also morphogenetic since cross talk between vessels, that is, endothelial cells, basement membrane compartments, and morphogenetic gene products bound to the vessel's basement membrane, cells, and surrounding extracellular matrices, triggers the ripple-like cascade of sequential inductive and differentiating events, leading to tissue morphogenesis and, on exposed dentine, cementogenesis with functionally inserted Sharpey's fibers. The morphogenetic vessels of Aristotelian definition (Crivellato et al. 2007) continuously provide progenitor/pericytic cells that ride each individual principal fiber, providing a continuous flow of progenitor cells to both the cementum and the alveolar bony side (Ripamonti et al. 2009a); in this context, the invading and sprouting capillaries are also morphogenetic, initiating the induction of alveolar bone and cementum. Expression of different phenotypes during the ride of progenitor cells along individual principal fibers, that is, cementoblastic or osteoblastic, is dependent on whether the common progenitor lineage rides closer to the cementum or the alveolar bony side of the periodontal ligament space. The final phenotype, that is, cementoblastic or osteoblastic, is acquired by riding into morphogen gradients of the extracellular matrix components of the alveolar bone or the exposed dentine and the newly deposited cementoid matrix (Ripamonti et al. 2009a).

The critical role of angiogenetic vessels and angiogenesis in periodontal tissue engineering is shown in Figure 7.5. Coronally to the osteogenetic front, capillary sprouting and invasion dictate the morphogenetic pattern of the induction of bone formation, and thus the supramolecular assembly of the osteonic remodeling bone (Figure 7.5a), the osteosome, the essential quantum structure of the primate remodeling osteonic bone (Reddi 1997). Figures 7.3 through 7.5 show angiogenesis, capillary sprouting, cell trafficking, and stem cell riding from the vascular to the alveolar bone and cementum compartments of the periodontal ligament space. Single invading capillaries dictate the morphogenetic pattern of the induction of bone formation (Figure 7.5). Mesenchymal cell condensations generate osteonic-like constructs around single central blood vessels, predating the osteonic central blood vessels of the remodeling osteosome (Figure 7.5) (Reddi 1997). High-power views show mesenchymal condensations populated by differentiating osteoblastic cells enveloping central blood vessels while generating the primordial osteonic bone centered on osteogenetic and morphogenetic blood vessels (Figure 7.5).

Sir Arthur Keith (1927) proposed that progenitor osteoblastic cell lines are endothelial cells derived by the invading osteogenetic and morphogenetic vessels. Levander (1938) further reported that "the impression is given that the fully formed mesenchymal cells ultimately emanate from the endothelial cells of the capillaries" (Levander 1938). In addition to endothelial cells, the capillary or microvessel pericytes may also be osteoblastic precursor cells (Brighton et al. 1992). Pericytes are elongated cells sharing a common basement membrane with

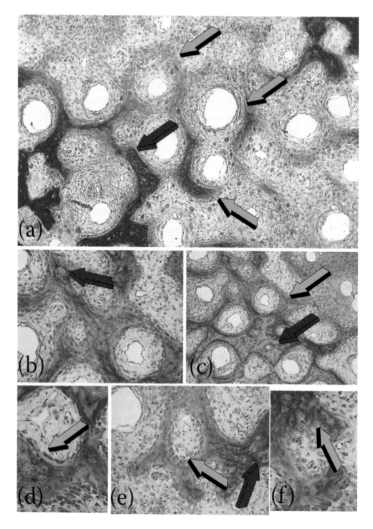

**FIGURE 7.5** Angiogenesis and the induction of bone formation by morphogenetic and osteogenetic vessels of Aristotle's (Crivellato et al. 2007) and Trueta's (1963) definitions. Vascular invasion and capillary sprouting provide the molecular, cellular, and morphological templates for the induction of bone formation by highly purified, naturally derived bone morphogenetic proteins (BMPs) implanted in Class II furcation defects of the non-human primate *Papio ursinus*. Single invading capillaries, in the context the osteogenetic vessels of Trueta's (1963) description and definition, dictate the pattern of the induction of bone formation and act as a template for the induction of the Haversian primate osteonic bone. The invading vessels are also morphogenetic since they initiate the induction of mesenchymal cellular condensations around each invading capillary. (a) Detail of supranotch Class II furcation defect treated with highly purified, naturally derived BMPs (Ripamonti et al. 1994). Invading osteogenetic vessels engineer cellular differentiation and mesenchymal condensations (light blue arrows) developing around each single capillary that controls the differentiation and development of lamellar

overlying endothelial cells. Pericytes encircle the endothelial cells, adding anatomical and functional plasticity, as well as stability, to capillary networks (Benjamin et al. 1998). Pericytes may differentiate into osteprogenitor cells as hypothesized since the classic studies on the bone induction principle (Urist 1965; Urist et al. 1967, 1968). More recent studies have additionally hypothesized the existence of perivascular progenitor stem cells, and thus of potential perivascular stem cell niches, providing an unlimited stem cell supply for tissue induction and regeneration (Benjamin et al. 1998; Zheng et al. 2007; Kovacic and Boehm 2009). Studies have now conclusively reported that perivascular multilineage progenitor cells (Chen et al. 2009) and multiorgan mesenchymal stem cells are the pericytes, indicating a perivascular origin for mesenchymal stem cells in multiple organs (Crisan et al. 2008).

The exquisite relationship between endothelial and pericytic cells of the vascular compartment and the periodontal ligament fibers, but particularly with the newly formed bone (Figure 7.5d–f), further indicates the tight homeostatic control of the periodontal ligament space. Endothelial cells contain microparticles (MPs) that are small membrane vesicles that can be secreted and released *via* an exocytotic budding process into the extracellular matrix (Lozito and Tuan 2011). After secretion, several matrix-altering proteases, including matrix metalloproteinases, are then released from the secreted MPs, thus interacting with the surrounding extracellular matrices modulating the homeostasis of vascular matrix remodeling (Lozito and Tuan 2011). Further induction and morphogenesis result in the induction of periodontal tissue regeneration with cementogenesis along the exposed root surfaces (Figure 7.6).

The tight morphological and thus molecular relationships between aggregated cellular elements with condensed chromatin interpreted as angioblasts (Figure 7.3b and c) (Ripamonti 2007; Ripamonti et al. 2009a), with newly generated Sharpey's fibers, indicate how precursor progenitor cells ride generated Sharpey's fibers long before the emergence of morphologically assembled capillaries within the periodontal ligament space. Angiogenesis

**FIGURE 7.5** *Continued*

osteonic bone. (b, c) Newly developed mesenchymal cellular condensation embraces each invading capillary, differentiating osteoblastic-like cells (light blue arrow) facing the central osteogenetic vessels, now controlling the genesis of the corticocancellous osteonic primate bone. Mesenchymal condensations mineralize (dark blue arrows), thus patterning the osteonic bone covered by plump osteoblast-like cells facing the central morphogenetic osteogenetic vessels (b–e).

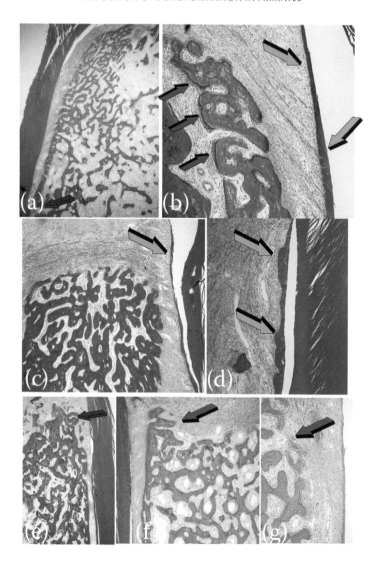

**FIGURE 7.6**  Tissue induction and morphogenesis of the complex morphologies of the periodontal tissues by highly purified, naturally derived bone morphogenetic proteins (BMPs) purified greater than 50,000-fold from crude extract after sequential chromatography on hydroxyapatite Ultrogel, heparin–Sepharose, and Sephacryl S-200 adsorption, affinity, and gel filtration chromatography (Luyten et al. 1989; Ripamonti et al. 1992). (a) Prominent periodontal tissue regeneration across the Class II furcation defect with substantial alveolar bone regeneration from the notched instrumented root surface (dark blue arrow). (b, c) Details of alveolar bone induction with newly formed mineralized bone in blue surfaced by as yet to be mineralized osteoid seams (magenta arrows in b) covered by contiguous osteoblasts secreting bone matrix. Mineralized cementum extends coronally along the instrumented root surfaces (light blue arrows in b, c). (d) Digital microphotographic detail of newly deposited mineralized cementum covered

is a prerequisite for osteogenesis (Trueta 1963); of interest, the same amino acid motifs embedded in the carboxy-terminal domain of osteogenic protein-1, also known as bone morphogenetic protein-7 (BMP-7), induce both angiogenesis (Ramoshebi and Ripamonti 2000) and osteogenesis (Reddi 2000; Ripamonti 2006). Osteogenic protein-1 is indeed at the crux of the complex cellular and molecular signals that regulate the topography and assembly of the extracellular matrix, precisely guiding angiogenesis, vascular invasion, osteogenesis, and cementogenesis with the induction of periodontal tissue regeneration (Ripamonti 2007). Osteogenic protein-1, unique among bone morphogenetic proteins, is both angiogenic and osteogenic, being the soluble molecular signal to initiate angiogenesis and osteogenesis simultaneously or, as previously reported, osteogenesis in angiogenesis (Ripamonti et al. 1996, 2006; Ramoshebi and Ripamonti 2000; Ripamonti 2007; Ripamonti et al. 2007).

While sprouting capillaries, endothelial and pericytic cells, and the extracellular matrix of the vascular compartment bound to angiogenic and bone morphogenetic proteins control and modulate the surrounding extracellular matrix, particularly bone homeostasis, little is known on the direct effect of the bone compartment on endothelial cells. The effects of BMPs/OPs have been studied on several cell types *in vitro* (Reddi 2000; Ripamonti 2006). Because of the intimate relationship between endothelium and osteoblasts, we have examined the effect of highly purified, naturally derived bone morphogenetic proteins on endothelial cells *in vitro* (Heliotis and Ripamonti 1994). Highly purified osteogenic fractions isolated from baboon diaphyseal bone matrix (Ripamonti et al. 1992) were added to cultures of aortic endothelial (E8) and rat vascular smooth muscle cells (RVSMCs), isolated and characterized as described (Jaffe et al. 1973; Howard et al. 1991).

A dramatic change in the morphology of E8 cells was observed with the addition of osteogenic fractions at concentrations of 3, 6, and 10 μg over 24 and 48 h; there was a change of cell morphology from a cobblestone to a spindle-shaped phenotype after 24 and 48 h (Figure 7.7), with eventual rounding up and detachment of cells by 72 h, regardless of the protein

FIGURE 7.6 *Continued*

by cementoid matrix surfaced by cementoblasts secreting cementoid matrix (light blue arrows). (e–g) Digital images of *de novo* induction of mineralized alveolar bone formation (dark blue arrow in e) covered by osteoid seams (magenta arrows in f, g) populated by contiguous osteoblasts. Capillary invasion and morphogenesis of the osteonic primate lamellar bone (details in f, g).

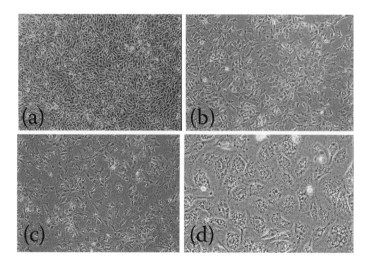

**FIGURE 7.7** Response of aortic endothelial (E8) cells to highly purified, naturally derived bone morphogenetic protein (BMP) fractions *in vitro* (Ripamonti et al. 1992; Heliotis and Ripamonti 1994). (a) E8 cells with cobblestone morphology prior to the addition of osteogenic proteins fractions. (b, c) E8 cells 48 h after the addition of osteogenic fractions at a concentration of 6 µg of protein in 300 µl of medium. Note the morphologic change to fibroblastic-like cells. (d) Further phenotypic changes 72 h after addition of osteogenic proteins with the differentiation of large hyperchromatic cells. Of note, reacquisition of the typical cobblestone appearance could be achieved when cultures were not fed osteogenic fractions for more than 48 h, providing that osteogenic fractions did not exceed a concentration of 6 µg of protein per 300 µl of medium (Heliotis and Ripamonti 1994). E8 endothelial cells did regain cobblestone morphology after 72 h without further addition of osteogenic protein fractions (Heliotis and Ripamonti 1994).

concentration used (Figure 7.7c and d) (Heliotis and Ripamonti 1994). Of note, the higher the protein concentration added, the earlier the changes were observed (Heliotis and Ripamonti 1994). Reacquisition of the cobblestone appearance could be achieved when cultures were not fed osteogenic fractions for more than 48 h, and provided that osteogenic fractions did not exceed a concentration of 6 µg of protein/300 µl of Dulbecco's modified Eagle's medium. Identical concentrations of osteogenic fractions did not alter the morphology of RVSMCs over 24, 48, and 72 h (Heliotis and Ripamonti 1994). Osteogenic fractions thus profoundly alter endothelial cell morphology *in vitro*, indicating an important role for bone matrix molecules on the phenotypic modulation of endothelial cells (Heliotis and Ripamonti 1994).

It is tempting to suggest that the observed profound phenotypic change in fibroblast spindle-like-shaped cells by osteogenic fractions, with further phenotypic changes 72 h after addition of osteogenic fractions with the differentiation of large hyperchromatic cells (Figure 7.7d), is the result of dedifferentiation of endothelial cells from the cobblestone phenotype in confluence by still unknown morphogenetic messages contained within highly purified osteogenic fractions *in vitro*. The binding of both angiogenic and osteogenic proteins to the extracellular matrix components of the sprouting capillaries is thus the supramolecular assembly of the morphogenetic and osteogenetic vessels of Aristotle's and Trueta's definition (Crivellato et al. 2007; Trueta 1963), assembling viable morphogenetic proteins in contact with the extracellular matrix for continuous differentiation or dedifferentiation of endothelial cells and pericytes into perivascular stemlike cells for the differentiation of rapidly dividing osteoblastic-like cells.

The pleiotropic activity of the bone morphogenetic proteins is vast and spans from neurogenesis to angiogenesis, from cardiogenesis to dentinogenesis, and from tooth morphogenesis with Sharpey's fibers functionally inserted into newly formed cementum to cerebellar patterning and differentiation (Thomadakis et al. 1999; Ripamonti 2006, 2007). The expression pattern of different BMPs during embryonic development further suggests novel strategies of therapeutic intervention outside the mere osseous domain (Vainio et al. 1993; Hogan 1996; Thesleff and Nieminen 1996; Thesleff and Sharpe 1997; Åberg et al. 1997; Thomadakis et al. 1999; Ripamonti 2006, 2007). These complex and vast pleiotropic activities of BMPs/OPs spring from minor amino acid sequence variations in the carboxy-terminal region of each protein isoform. Indeed, BMPs/OPs elegantly reflect Nature's parsimony in controlling multiple specialized functions or pleiotropy, deploying molecular isoforms with minor variations in amino acid sequence motifs within highly conserved carboxy-terminal regions to control multiple specialized functions or pleiotropy (Thomadakis et al. 1999; Reddi 2000; Ripamonti et al. 2001; Ripamonti 2007; Ripamonti and Petit 2009). Such amino acid variation confers specialized activities to each BMP/OP isoform, the molecular basis that determines the structure/activity profile of the bone morphogenetic proteins (Ripamonti and Reddi 1994; Thomadakis et al. 1999; Ripamonti et al. 2001, 2005; Ripamonti 2007; Ripamonti and Petit 2009).

The biological significance of apparent redundancy and its therapeutic implications rest on developing the structure/activity profile of the members of the BMP/OP family, that is, to study in vivo and in primate tissues the morphogenetic impulse of each single and structurally different recombinant protein so

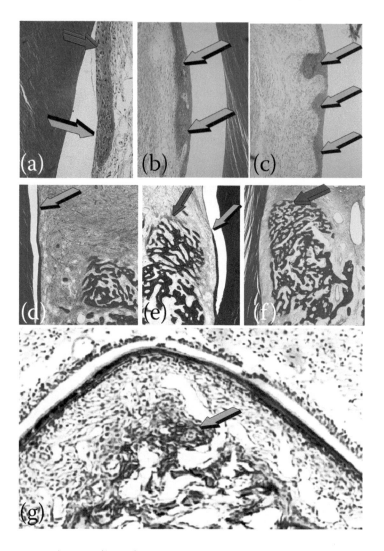

**FIGURE 7.8** Induction of bone formation as a recapitulation of embryonic development. Pleiotropic activity and structure/activity profile of bone morphogenetic proteins (BMPs) in tissue induction and morphogenesis of the specialized morphologies of the periodontal tissues in the non-human primate Chacma baboon (*Papio ursinus*). (a–c) Digital microphotographs at different apicocoronal levels of cementogenesis as induced by doses of recombinant human osteogenic protein-1 (hOP-1) along planed and exposed dentinal surfaces. In context of periodontal tissue regeneration and in contact with the exposed dentine matrix, hOP-1 is preferentially cementogenic when implanted in Class II furcation defects. Induction of cementum with cementoid matrix deposition (magenta arrows in a) and hypercellularity with nascent mineralization (light blue arrow) in the as yet to be fully mineralized cementum characterize tissue induction and morphogenesis by hOP-1 when implanted in furcation defects of the Chacma baboon. (b, c) Prominent cementogenesis by hOP-1 along denuded root surfaces (light blue arrows in b and c)

as to discover pleiotropic activities of molecularly similar, but structurally different, isoforms based on specific amino acid sequence motifs. Indeed, osteogenic protein-1 provides important morphological and molecular evidence of its pleiotropic activity in tissue induction and morphogenesis (Figure 7.8) (Ripamonti et al. 2001, 2009a; Ripamonti 2006, 2007).

In the context of periodontal tissue regeneration and in contact with dentine extracellular matrices, osteogenic protein-1, at the doses tested in furcation defects of the Chacma baboon (*P. ursinus*), is preferentially, if not strictly, cementogenic (Figure 7.8a–c) (Ripamonti et al. 1996, 2009a; Ripamonti 2006, 2007). Cementogenesis is not confined to only along the exposed root surfaces, but extends to the induced mineralized tissue surrounding the insoluble collagenous bone matrix as carrier within the treated furcation defects (Ripamonti et al. 1996; Ripamonti and Reddi 1997; Ripamonti 2007). Cementum induced within the furcation defects is attached to the root dentine, and the newly formed cementum is separated from the remaining alveolar bone by a pseudoligament space, with fibers originating from both the cemental and the alveolar bone interfaces (Ripamonti et al. 1996; Ripamonti 2007). When implanted in either surgically created or inflammatory-induced furcation defects of *P. ursinus*, hOP-1 is highly cementogenic (Figure 7.8c and d) (Ripamonti et al. 1996, 2002; Ripamonti 2007), as the protein does enhance cementoblast function *in vitro* (Hakki et al. 2010). The preferential induction of cementogenesis by hOP-1 when in contact with dentine extracellular matrices is further supported by the proteome and gene expression profile of cementoblasts treated by hOP-1 *in vitro* (Bozic et al. 2012).

**FIGURE 7.8** *Continued*

with hypercellularity and predictable deposition of cemental matrix by active cementoblasts inducing substantial cementogenesis. (d) In different experiments, cementogenesis by 100 μg of hOP-1 (light blue arrow). (e) Induction of newly formed and mineralized bone covered by osteoid seams (magenta arrow) in furcation defect treated with 100 μg of human recombinant bone morphogenetic protein-2 (hBMP-2) (Ripamonti et al. 2001). In contrast, limited cementogenesis along the exposed root surface by hBMP-2 (light blue arrow in e). Restoration of the induction of cementogenesis and alveolar bone regeneration with large osteoid seams surfacing newly formed mineralized bone (magenta arrow) by binary application of 100 μg of hOP-1 and 100 μg of hBMP-2 60 days after implantation in Class II furcation defects of *P. ursinus* (Ripamonti et al. 2001). (e) Mechanistic insights into different qualitative and quantitative responses by molecularly different but homologous BMP isoforms during induction of periodontal tissue morphogenesis. Immunolocalization of BMP-2 during tooth morphogenesis of a mandibular molar of a 16-day-old mouse pup (Thomadakis et al. 1999) is strictly localized in the developing alveolar bone only (light blue arrow) during tooth morphogenesis.

In marked contrast, bone morphogenetic protein-2 (hBMP-2) is highly osteogenic when implanted in periodontal defects of a variety of animal models, including canines and non-human primates (Sigurdsson et al. 1995a, 1995b; Giannobile et al. 1998; Choi et al. 2000; Ripamonti et al. 2001; Ripamonti 2007). In Class II furcation defects of *P. ursinus*, hBMP-2 is preferentially osteogenic and shows limited cementogenesis (Figure 7.8e). This is also found in canine models (Sigurdsson et al. 1995a, 1995b; Giannobile et al. 1998; Choi et al. 2000). Using hBMP-2 in surgically induced three-wall intrabony periodontal defects in young adult mongrel dogs, Choi et al. (2000) found cementum regeneration to be moderate, confirming previous published results obtained in canine (Sigurdsson et al. 1995a, 1995b; Giannobile et al. 1998; Choi et al. 2000) and non-human primate models (Ripamonti et al. 2001).

Mechanistically, the limited induction of cementogenesis by hBMP-2 is explained by the reported data that hBMP-2 inhibits differentiation and mineralization of cementoblasts in vitro (Zhao et al. 2003). Exposure of cementoblasts to BMP-2 in vitro results in dose-dependent reduction of bone sialoprotein and collagen type I gene expression, ultimately inhibiting cell-induced mineral nodule formation (Zhao et al. 2003). The specificity of osteogenic protein-1 primarily initiating cementogenesis when implanted in periodontal defects of canine and non-human primate models is regulated by both the dentine extracellular matrix upon which responding cells differentiate and the specific structure/activity profile of the implanted recombinant morphogen (Ripamonti et al. 1996; Ripamonti 2007). Importantly, the biological activity of naturally derived and recombinant hBMPs can be modulated and selectively potentiated by binding affinities for different extracellular matrix substrata; this underscores the critical regulatory role of extracellular matrices in the phenotypic modulation of a variety of cells (Reddi 1997, 2000).

An important question for tissue engineering and regenerative medicine at large is whether the presence of molecularly different osteogenic proteins of the TGF-β superfamily, that is, the BMPs/OPs and the mammalian TGF-β isoforms in primate only, has a therapeutic significance (Ripamonti and Reddi 1994; Ripamonti 2007; Ripamonti and Petit 2009). Several studies in canine and non-human primate models have unequivocally shown that hOP-1 combined either with bovine or baboon collagenous bone matrices or with a growth factor–reduced Matrigel matrix preferentially induces cementogenesis as evaluated on undecalcified sections (Ripamonti et al. 1996, 2001, 2002; Ripamonti 2007). The induction of cementogenesis along the exposed root surfaces is substantial, with the induction of the highly cellular cementoid matrix with foci of mineralization well before the generation of Sharpey's fibers inserting into the newly formed cementum (Figure 7.8a–c).

Furcation defects prepared in the first and second mandibular molars of adult Chacma baboons were used to assess whether tissue induction could be enhanced and tissue morphogenesis modified by binary or single applications of hOP-1 and hBMP-2. As reported in previous experiments (Ripamonti et al. 1996), 100 µg doses of hOP-1 implanted in mandibular molar furcation defects harvested 60 days after operation showed substantial cementogenesis (Figure 7.8c and d) (Ripamonti et al. 2001). On the other hand, 100 µg doses of hBMP-2 induced greater amounts of mineralized bone and osteoid than hOP-1-treated defects with limited induction of cementogenesis (Figure 7.8e). The results of the study, which is the first to attempt to address the structure/activity profile among BMP/OP family members in vivo, have indicated that periodontal tissue morphogenesis induced by hOP-1 and hBMP-2 is qualitatively different when the morphogens are applied singly, with hOP-1 inducing substantial cementogenesis; on the other hand, hBMP-2-treated defects showed limited cementum formation, but a temporal enhancement of alveolar bone regeneration and remodeling (Ripamonti et al. 2001).

Importantly, immunolocalization of BMP-2 is strictly confined to the alveolar bone during tooth morphogenesis of mouse pups (Figure 7.8g) (Thomadakis et al. 1999), further indicating the primary functional characteristic of the molecular isoform following the basic principle that gene products exploited in embryonic development are redeployed postnatally to induce tissue induction and morphogenesis of the same tissues assembled in development (Ripamonti 2003, 2006, 2007).

## 7.5 Induction of Periodontal Tissue Regeneration by Recombinant Human TGF-$\beta_3$ in *P. ursinus*

In the non-human primate *P. ursinus*, in marked contrast to rodents, lagomorphs, and canines, the heterotopic induction of bone formation is not limited to the bone morphogenetic proteins, but extends to molecularly related but different molecular isoforms of the TGF-$\beta$ supergene family, that is, the three mammalian TGF-$\beta$ isoforms (Figure 7.2) (Ripamonti et al. 1997,

2000b, 2008; Ripamonti and Roden 2010). The induction of bone formation is not limited to extraskeletal *rectus abdominis* heterotopic sites, but encompasses significant amounts of alveolar bone regeneration with embedded Sharpey's fibers uniting the alveolar bone to the newly formed cementum (Figure 7.9) (Teare et al. 2008; Ripamonti et al. 2009a, 2009b). Two novel and provocative treatments recently performed in *P. ursinus* using the recombinant hTGF-$\beta_3$ induced substantial periodontal tissue regeneration in Class II and III furcation defects of the first and second mandibular molars (Teare et al. 2008; Ripamonti et al. 2009a, 2009b).

The direct application of hTGF-$\beta_3$ in a Matrigel matrix with the addition of *rectus abdominis* muscle cells *versus* the transplantation of fragments of heterotopic ossicles induced by different doses of the hTGF-$\beta_3$ protein resulted in highly comparable periodontal tissue regeneration in Class II furcation defects of *P. ursinus* (Teare et al. 2008; Ripamonti et al. 2009a, 2009b). The rapid architectural sculpture of mineralized corticalized constructs in the *rectus abdominis* muscle by the hTGF-$\beta_3$ isoform of *P. ursinus* (Figure 7.2) is a novel source of developing autoinduced bone for autogenous transplantation in preclinical and clinical contexts (Ripamonti et al. 2008) as tested in Class II and III furcation defects of the Chacma baboon (*P. ursinus*) (Teare et al. 2008; Ripamonti et al. 2009a, 2009b).

Additional studies were thus designed to establish whether the direct application of hTGF-$\beta_3$, together with responding *rectus abdominis* muscle cells, would result in superior periodontal tissue regeneration. A series of experiments were thus performed in mandibular Class II and III furcation defects of *P. ursinus* (Teare et al. 2008; Ripamonti et al. 2009a, 2009b). The studies revealed that 75 μg of hTGF-$\beta_3$ combined with growth factor–reduced Matrigel matrix results in significant osteogenesis, together with cementogenesis, along the exposed root surfaces (Figure 7.9). Of interest, single applications of hOP-1 and hTGF-$\beta_3$ generated similar amounts of mineralized bone within treated furcation defects. Of note, hTGF-$\beta_3$ in Matrigel matrix induced a substantial amount of cementogenesis along the exposed root surfaces, with thick deposits of cementoid and mineralized cementum (Figure 7.9d).

High-power views of engineered periodontal tissues by the hTGF-$\beta_3$ osteogenic device in growth factor–reduced Matrigel matrix highlight the induction of Sharpey's fibers with angiogenesis regulating the assembly of the inserting fibers (Figure 7.9a and b) spanning across the periodontal ligament system, anchoring onto the newly formed cementum

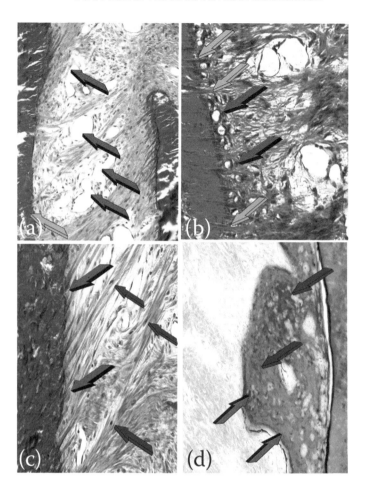

**FIGURE 7.9** Induction of cementogenesis, Sharpey's fibers, angiogenesis and capillary sprouting, cellular trafficking, and the induction of bone formation by 75 μg of recombinant human transforming growth factor-$\beta_3$ (hTGF-$\beta_3$) in growth factor–reduced Matrigel matrix in Class II mandibular molar furcation defects of the Chacma baboon (*Papio ursinus*). (a, b) *De novo* insertion of *bona fide* Sharpey's fibers within the exposed dentine matrix across the newly formed cementum covered by layered cementoblasts secreting cementoid matrix as yet to be mineralized. (b) Riding progenitor cells along single individual periodontal ligament fibers (dark blue arrows in b) with cementoblasts stacked between Sharpey's fibers along the newly formed and mineralized cementum (magenta arrows in b). (c) Prominent angiogenesis and capillary sprouting surrounded by hyperchromatic cellular progenitors and pericytes expanding across the tridimensional network of the periodontal ligament system providing a continuous flow of responding stem cells (dark blue arrows) to the cementoid and osteoid sides of the newly induced periodontal ligament system. (d) High-power view of the induction of cementoid deposition (magenta arrows) with foci of nascent mineralization (dark blue arrows) across the newly formed remodeling cementoid matrix; significant induction of cementogenesis by the third mammalian transforming growth factor-$\beta$ isoform.

(Figure 7.9a and b). Capillary sprouting and invasion always characterize hTGF-$\beta_3$-treated specimens with periodontal ligament fibers inserting into newly formed cementum and into newly formed bone as nonmineralized fibers into an osteoid matrix populated by contiguous osteoblasts (Figure 7.9a).

Tissue engineering using pluripotent stem cells comes with the challenges of tissue compatibility and the ethical dilemma of the use of human embryos. Thus, the focus of regenerative medicine has turned to the potential use of postnatal stem cells, such as muscle-derived striated cells (MDSCs) (Usas et al. 2011), and particularly pluripotent stem cells induced by defined factors (Takahashi and Yamanaka 2006; Takahashi et al. 2007). Several studies undertaken in *P. ursinus* have shown the critical role of skeletal muscle–derived cells in bone induction and morphogenesis (Ripamonti et al. 2008, 2009a, 2009b; Teare et al. 2008). Morcellated fragments of autogenous rectus abdominis muscle partially restored the endochondral osteoinductivity of the hTGF-$\beta_3$ osteogenic device when implanted in calvarial defects of adult baboons (Ripamonti et al. 2008, 2009c).

Experiments were thus designed to study the effect of morcellated fragments of autogenous *rectus abdominis* muscle when implanted in Class II and III furcation defects of adult baboons. Harvested fragments of autogenous rectus abdominis muscle were finely minced and thoroughly morcellated. Morcellated muscle tissue was added to 75 µg of hTGF-$\beta_3$ in Matrigel matrix and implanted in furcation defects in *P. ursinus* (Teare et al. 2008; Ripamonti et al. 2009a, 2009b). The direct application of 75 µg of hTGF-$\beta_3$ in a Matrigel matrix, together with the addition of finely minced fragments of autogenous *rectus abdominis* muscle, resulted in greater alveolar bone formation and cementogenesis than periodontal defects treated by hTGF-$\beta_3$ alone in Matrigel matrix (Figure 7.10) (Teare et al. 2008; Ripamonti et al. 2009a, 2009b). The addition of responding stem cells prepared by finely morcellating fragments of autogenous *rectus abdominis* muscle significantly enhanced the induction of periodontal tissue regeneration. Importantly, myoblastic/pericytic stem cells prepared by finely mincing fragments of autogenous *rectus abdominis* muscle significantly increase the coronal extent of cementogenesis along surgically exposed root surfaces (Figure 7.10) (Ripamonti et al. 2009a, 2009b).

By simply morcellating fragments of autogenous *rectus abdominis* biopsies, we have shown that the striated *rectus abdominis* muscle is an important source of myoblastic/perivascular and pericytic stem cells that can be rapidly prepared and transplanted in periodontal defects of the non-human primate

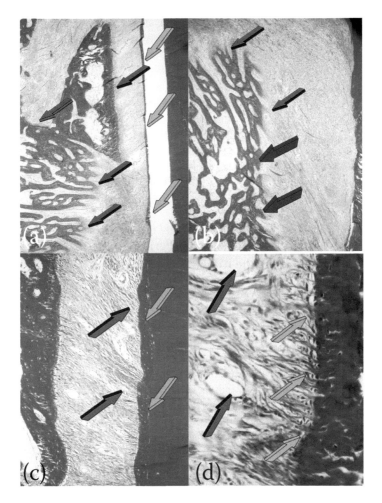

**FIGURE 7.10**    Periodontal tissue induction and regeneration by 75 µg of recombinant human transforming growth factor-$\beta_3$ (hTGF-$\beta_3$) in growth factor–reduced Matrigel matrix in Class II mandibular molar furcation defects of the Chacma baboon (*Papio ursinus*). Furcation defects were supplemented by morcellated fragments of autogenous *rectus abdominis* muscle co-implanted within the Matrigel matrix. (a) Substantial cementogenesis along the exposed root surface (light blue arrows) face newly formed mineralized bone covered by osteoid seams (magenta arrows) populated by contiguous osteoblasts. (b) Detail of instrumented root surface with induction of alveolar bone (dark blue arrows) covered by large osteoid seams (magenta arrows). (c) Thick mineralized cementum deposition and remodeling (light blue arrows) with inserted Sharpey's fibers (magenta arrows). (d) Detail of more coronal tissue patterning by the hTGF-$\beta_3$ isoform showing *de novo*–generated Sharpey's fibers from the exposed dentine matrix almost touching newly formed capillaries (magenta arrows). Alignment of newly induced Sharpey's fibers provides the extracellular matrix scaffolding to stack cementoblastic cells between fibers (light blue arrows), and thus the continuous availability and secretion of cementoid matrix from the available cementoblasts.

Chacma baboon (Teare et al. 2008; Ripamonti et al. 2009a, 2009b). Importantly, the above results using seemingly crude preparation of cellular material contained in morcellated fragments of autogenous *rectus abdominis* striated muscle have indicated that the striated muscle retains responding mesenchymal stem cells capable of transformation into desired cellular phenotypes, that is, osteoblastic and cementoblastic cell lines, respectively, when in contact with specific extracellular matrix substrata, thus engineering calvarial (Ripamonti et al. 2008, 2009c) and periodontal tissue induction (Teare et al. 2008; Ripamonti et al. 2009a, 2009b).

## 7.6 Challenges and Perspectives in Regenerative Medicine of the Periodontal Tissues by the hTGF-$\beta_3$ Isoform

The antiquity and severity of periodontal diseases have been provided by the fossilized hard evidence of alveolar bone loss in Pliocene and early Pleistocene gnathic remains unearthed at the Blaauwbank Valley in Sterkfontein and Swartkrans, South Africa (Figure 7.1) (Ripamonti 1988, 1989). Several million years after the Australopithecinae and early Homo taxa suffered from alveolar bone loss with *de facto* concomitant gingivitis and periodontitis, extant Homo identified the vast pleiotropic activities of members of the TGF-$\beta$ supergene family (Vainio et al. 1993; Hogan 1996; Thesleff and Nieminen 1996; Thesleff and Sharpe 1997; Åberg et al. 1997; Thomadakis et al. 1999; Ripamonti 2006, 2007). Selected gene products of the TGF-$\beta$ supergene family also initiate the induction of periodontal tissue regeneration with the induction of cementogenesis as a function of the structure/activity profile of each recombinant human protein (Ripamonti and Reddi 1994, 1997; Thomadakis et al. 1999; Ripamonti 2007).

A number of studies have shown that partially purified cemental proteins contain mitogenic proliferative and attachment proteins interacting with cementum extracellular matrix selectively affecting periodontal cell populations within the periodontal ligament space (Nakae et al. 1991; Yonemura et al. 1992; Narayanan and Yonemura 1993; Pitaru et al. 1993; Wu et al. 1996; Liu et al. 1997). In addition to cementum-derived attachment proteins potentially related to the development of the cementoblastic phenotype, the avascular cemental matrix may contain osteogenic proteins involved in self-repair and self-inductive phenomena interacting with both cementoblasts and periodontal ligament cells to maintain and self-repair the functionally oriented collagenous fibers of the periodontal ligament system. To further understand the multiple effects of cemental homeostasis on the periodontal ligament space, an important challenge is to finally address the inductive activity of cemental extracts after chaotropic dissociative extraction and heterotopic implantation of purified cemental proteins.

Mechanistically, the specificity of hOP-1 primarily initiating cementogenesis when implanted in periodontal furcation defects of canine and non-human primate models is regulated by the dentine extracellular matrix, upon which responding cells attach and differentiate, and by the specific structure/activity profile of the implanted molecular isoform (Ripamonti et al. 1996; Ripamonti and Reddi 1997; Ripamonti 2007; Ripamonti and Petit 2009). The biological activity of naturally derived and recombinant hBMPs/OPs can be modulated and selectively potentiated by binding affinities for different extracellular matrix substrata; this underscores the critical regulatory role of extracellular matrices in the phenotypic modulation of a variety of cells (Reddi 1997, 2000). The limited induction of cementogenesis by recombinant hBMP-2 is additionally explained by the reported data that hBMP-2 inhibits differentiation and mineralization of cementoblasts (Zhao et al. 2003).

To further understand the multiple effects of cemental homeostasis and regenerative phenomena of the cementum and periodontal ligament space, it is now mandatory to study the inductive activity, if any, of cemental extracts after chaotropic dissociative extraction of the cemental extracellular matrix. Cemental protein extracts have shown that partially purified proteins are regulators of chemoattraction and cell differentiation, and thus critical during the process of periodontogenesis (Nakae et al. 1991; Yonemura et al. 1992; Narayanan and Yonemura 1993; Pitaru et al. 1993; Wu et al. 1996; Liu et al. 1997). Osteogenic proteins of the TGF-β superfamily may be present within the cemental matrix as a memory of developmental events during the induction of cementogenesis and the assembly of the periodontal ligament space (Moehl and Ripamonti 1992; Ripamonti and Renton 2006).

The concept of osteogenic proteins embedded within the cemental matrix as a memory of developmental events has been highlighted by the induction of bone formation by demineralized allogeneic dentin matrices when implanted in the *rectus abdominis* muscle of *P. ursinus* (Moehl and Ripamonti 1992). The findings of mineralized matrices of bone/cementum-like deposits after transplanting human cementum–derived cells in immunodeficient mice (Grzesik et al. 1998) further indicate that cementum may contain embryonic remnants of osteogenic proteins that were required during dentinogenesis and cementogenesis in embryonic tooth development (Vainio et al. 1993; Hogan 1996; Thesleff and Nieminen 1996; Thesleff and Sharpe 1997; Åberg et al. 1997; Thomadakis et al. 1999; Moehl and Ripamonti 1992).

The extracellular matrix of the cementum containing growth and mitogenic factors regulates the adjacent periodontal ligament space and provides the supramolecular framework for the regulation of the various components of the periodontal ligament sequestering and releasing morphogenetic factors involved in repair, regeneration, and remodeling of the periodontal ligament space (Moehl and Ripamonti 1992; Åberg et al. 1997; Thomadakis et al. 1999; Ripamonti 2007).

Ultimately, it is now necessary to unravel the distinct spatial and temporal patterns of the molecular and cellular cascades regulating cementogenesis with *de novo* generated Sharpey's fibers tightly locked into the newly formed cemental matrix. Our continuous studies on the induction of bone formation by the hTGF-$\beta_3$ isoform have shown that the inductive and morphogenetic effects of hTGF-$\beta_3$ are set into motion and modulated by the expression of selected *bone morphogenetic proteins* (Ripamonti et al. 2014; Klar et al. 2014). Similarly, the induction of cementogenesis and alveolar bone regeneration by the hTGF-$\beta_3$ isoform in *P. ursinus* is invocated by the sequential and synchronous patterns of *BMP/OP* expression upon the implantation of the hTGF-$\beta_3$ isoform.

Striated muscle is known to harbor myoblast/myoendothelial and perivascular/pericytic stem cells capable of rapid differentiation into secreting osteoblasts (Zheng et al. 2007; Kovacic and Boehm, 2009). Class II and III furcation defects in *P. ursinus* were treated with 75 μg of hTGF-$\beta_3$ in a growth factor–reduced Matrigel matrix with morcellated fragments of autogenous *rectus abdominis* muscle (Ripamonti et al. 2009a, 2009b; Ripamonti and Petit 2009).

Of great interest, although stem cells, together with differentiated muscle cells, were delivered by fundamentally crude, morcellated *rectus abdominis* muscle preparations, comparative histomorphometrical analyses of morcellated fragments co-implanted with hTGF-$\beta_3$ osteogenic devices in a variety of different biological microenvironments induced greater cementogenesis and alveolar bone regeneration (Ripamonti et al. 2009a, 2009b), not the least of which is the induction of bone formation in non-healing calvarial defects of *P. ursinus* (Ripamonti et al. 2008, 2009c).

In spite of the abundance of studies at the morphological level, to date the molecular and cellular mechanisms that set *in vivo* postnatal induction of periodontal tissue regeneration and the initiation of cementogenesis are still unknown. The functionality of engineered periodontal tissues has been evaluated primarily by the morphological and morphometrical evidence of cementogenesis with functionally oriented periodontal ligament fibers (Bartold et al. 2000; Ripamonti 2007; Ripamonti et al. 2009a). We have already proposed that the induction of bone formation, together with the induction of cementogenesis, should be based on the expression patterns of selected gene products as a recapitulation of embryonic development (Ripamonti 2005, 2007; Ripamonti et al. 2009a).

Preliminary data in *Macaca mulatta* monkeys (Klar and Ripamonti 2013, Bone Research Laboratory, unpublished data) show by qRT-PCR molecular analyses that *osteogenic protein-1* is expressed in cemental matrices harvested following root planing of surgically exposed normal roots (Figure 7.11). Figure 7.11 shows gene-specific oligonucleotide sequences of the selected primers for cementum protein-1 (CEMP1) (Carmona-Rodriguez et al. 2007), osteocalcin (OC), transforming growth factor-$\beta_3$ (TGF-$\beta_3$), bone morphogenetic protein-2 (BMP-2), osteogenic protein-1 (OP-1), and collagen type IV (COL IV), the latter a marker of angiogenesis being the cementum avascular. The molecular data show that *OP-1* is expressed within the harvested normal cementum after root planing of surgically exposed but normal roots (Figure 7.11). *OP-1* is a critical regulatory gene involved in self-repair and self-inductive phenomena of both cementoblasts and periodontal ligament cells to induce and maintain cementogenesis with functionally inserted fibers (Amar et al. 1997; Ripamonti 2007; Hakki et al. 2010; Bozic et al. 2012). Of note, *OP-1* in *M. mulatta* cementum showed a twofold increase when compared to *BMP-2* (Figure 7.11) (Klar and Ripamonti 2013, Bone Research Laboratory, unpublished data). Importantly, the gene expression's pattern correlates with the morphological and morphometrical evidence of limited induction of cementogenesis by hBMP-2 when implanted in both canine and non-human primate models (Sigurdsson et al. 1995a, 1995b; Giannobile et al. 1998; Choi et al. 2000; Ripamonti et al. 2001). It was noteworthy that both *OP-1* and *BMP-2* are expressed and synthesized within the cementum in different ratios according to the structure/activity profile of the secreted proteins (Ripamonti 2007; Ripamonti and Petit

(a)

| Target gene | Forward primer (5'-3') | Reverse primer (5'-3') | NCBI Accession number |
|---|---|---|---|
| CEMP1 | AACCTCACCTGCCTCTCC | CTCCCTTCCCCCTTGCTTAC | NM 001048212 |
| OC | AGGGCAGCGAGGTAGTGAAG | CCTGAAAGCCGATGTGGT | BC113432 |
| TGF-β3 | AAACTCCAGCTCCCTTCC | TGCACTGCGAGAGCTTC | NM 003239.2 |
| BMP-2 | AACACTGTGCGCAGCTTG | TCCACCACCACACAAGCA | NG 023233.1 |
| OP-1 | TGTCACACAGGCTGGAGT | TCGCTTGAGGTCAGGACT | NM 001719.2 |
| COL IV | CAGTTGCCTTTGCATGCT | CAGCATGGGGGAAACTG | NG 011544.1 |
| GAPDH | CCACCCATGGCAAATTCC | TTCCCCCTGCAAGTGAAC | XR 094634.1 |

(b)

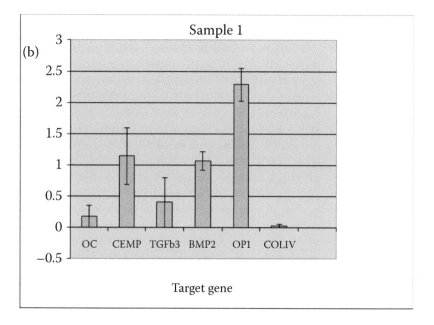

**FIGURE 7.11** Primer sequences used for qRT-PCR of target genes in baseline normal cementum planed and prepared from *Macaca mulatta* nonexposed roots. Sample 1: Relative gene expression in baseline root-planed cementum showing *CEMP1, BMP-2,* and substantial *OP-1* gene expression. CEMP1, cementum protein-1 (Carmona-Rodriguez et al. 2007); OC, osteocalcin; TGF-β3, transforming growth factor-β3; BMP-2, bone morphogenetic protein-2; OP-1, osteogenic protein-1; COL IV, collagen type IV; GAPDH, glyceraldehyde 3-phosphate dehydrogenase.

2009). Baseline data after root cementum instrumentation on nonexposed normal roots of *M. mulatta* mandibular molars also show the expression of *Cementum Protein-1*, a secreted gene product inducing the expression of bone and cementum proteins by human gingival fibroblast (Carmona-Rodriguez et al. 2007).

We have shown that the growth factor–reduced Matrigel matrix, together with hTGF-$\beta_3$, provides a morphogenetic regenerative microenvironment in which myoblastic/myoendothelial and perivascular/pericytic stem cells, released by the implanted morcellated fragments of the autogenous *rectus abdominis* muscle, migrate and attach to the dentinal substratum, eventually differentiating into secreting cementoblasts initiating cementogenesis on the exposed dentine. Invading and newly induced sprouting capillaries within the newly formed periodontal ligament space provide a niche of progenitor stem cells originating within the vascular compartment (Zheng et al. 2007; Kovacic and Boehm 2009; Ripamonti et al. 2009a). Pericytes and endothelial cells migrate from the vascular compartment into the periodontal ligament space by riding single periodontal ligament fibers toward the cementogenic or osteogenic sides of the periodontal ligament space, ultimately differentiating by morphogen gradients across the ride (Ripamonti et al. 2009a). The ride of progenitor stem cells across morphogen gradients along single periodontal ligament fibers is the basic supramolecular assembly for the maintenance and regeneration of the periodontal ligament space (Ripamonti et al. 2009a).

Recent studies on the mechanical regulation of cell function have shown the critical role of geometrically regulated substrata's rigidity in modulating cell signaling (Disher et al. 2005). Human mesenchymal stem cells (hMSCs) that underwent osteogenic differentiation demonstrated higher traction forces and cell spreading when regulated by the rigidity of the substratum (Fu et al. 2010). In the context of the periodontal wound model, it is tempting to suggest that cementogenic or osteogenic differentiation of locally available stem cells or precursors is additionally regulated by the rigidity of the available substrata. Highly rigid avascular mineralized dentinal substrata would induce a higher degree of cell spreading with highly organized actin stress fibers with large focal adhesions (Disher et al. 2005; Fu et al. 2010), characteristics of cementoblasts migrating and spreading on mineralized and planed dentinal surfaces. This would suggest cementoblasts endowed with prominent tractional forces for coronal migration along the planed, highly mineralized rigid dentinal substrata. Importantly, stem cells thus "feel the difference" between soft and hard substrata (Buxboim and Disher 2010) as a geometrically regulated cue integrated with the soluble molecular signal of the hTGF-$\beta_3$ protein that ultimately directs cell differentiation and the induction of cementogenesis along the exposed dentinal surfaces.

Finally, the real progress defining tissue engineering and regenerative medicine would be to define and control the microenvironmental niches within each tissue and perhaps also within different anatomical regions of a given tissue (Badylak and Nerem 2010). This will now be a fertile area of current and future research. It will be important to grasp whether soluble molecular signals, morphogenetic and differentiating gradients, pleiotropic and threshold activities, and ligand interactions and concentrations are likely to be substantially different at the cementum–periodontal ligament interface on the roots of different teeth, maxillary *versus* mandibular, and within segments of the roots, that is, buccal *versus* lingual, mesial versus distal, and apical *versus* coronal, so as to sustain different anatomical and functional microenvironments of each root and of each segment of the root in functional operation (Magan and Ripamonti 2013).

## References

ÅBERG, T., WOZNEY, J., THESLEFF I. (1997). Expression patterns of bone morphogenetic proteins (Bmps) in the developing mouse tooth suggest roles in morphogenesis and cell differentiation. *Dev Dyn* 210(4), 383–96.

AMAR, S., CHUNG, K.M., NAM, S.H., KARATZAS, S., MYOKAI, F., VAN DYKE, T.E. (1997). Markers of bone and cementum formation accumulate in tissues regenerated in periodontal defects treated with expanded polytetrafluoroethylene membranes. *J Periodontal Res* 32(1 Pt 2), 148–58.

BADYLAK, S.F., NEREM, R.M. (2010). Progress in tissue engineering and regenerative medicine. *Proc Natl Acad Sci USA* 107(8), 3285–86.

BARTOLD, P.M., MCCULLOCH, C.A., NARAYANAN, A.S., PITARU, S. (2000). Tissue engineering: A new paradigm for periodontal regeneration based on molecular and cell biology. *Periodontol 2000* 24, 253–69.

BENJAMIN, L.E., HEMO, I., KESHET E. (1998). A plasticity window for blood vessel remodeling is defined by pericytes coverage of the preformed endothelial network and is regulated by PDGF-B and VEGF. *Development* 125(9), 1591–98.

BOZIC, D., GRGUREVIC, L., ERJAVEC, I., BRKLJACIC, J., ORLIC, I., RAZDOROV, G., GRGUREVIC, I., VUKICEVIC, S., PLANCAK, D. (2012). The proteome and gene expression profile of cementoblastic cells treated by bone morphogenetic protein-7 in vitro. *J Clin Periodontol* 39(1), 80–90.

BRIGHTON, C.T., LORICH, D.G., KUPCHA, R., REILLY, T.M., JONES, A.R., WOODBURY, R.A. (1992). The pericyte as a possible osteoblast progenitor cell. *Clin Orthop Relat Res* 275, 287–99.

BUXBOIM, A., DISHER D.E. (2010). Stem cells feel the difference. *Nat Methods* 7(9), 695–97.

CARMONA-RODRIGUEZ, B., ALVAREZ-PÉREZ, M.A., NARAYANAN, A.S., ZEICHNER-DAVID, M., REYES-GASGA, J., MOLINA-GUARNEROS, J., GARCÍA-HERNÁNDEZ, A.L., SUÁREZ-FRANCO, J.L., CHAVARRÍA, I.G., VILLARREAL-RAMÍREZ, E., ARZATE, H. (2007). Human cementum protein 1 induces expression of bone and cementum proteins by human gingival fibroblasts. *Biochem Biophys Res Commun* 358(3), 763–69.

CHEN, C.W., MONTELATICI, E., CRISAN, M., CORSELLI, M., HUARD, J., LAZZARI, L., PÉAULT, B. (2009). Perivascular multi-lineage progenitor cells in human organs: Regenerative units, cytokine sources or both? *Cytokine Growth Factor Rev* 20(5–6), 429–34.

CHINSEMBU, K.C. (2012). Teeth are bones: Signature genes and molecules that underwrite odontogenesis. *J Med Genet Genom* 4(2), 13–24.

CHOI, S.-H., KIM, C.-K., CHO, K.-S., HUH, J.-S., SORENSEN, R.G., WOZNEY, J.M., WIKESJÖ, U.M.E. (2000). Effect of recombinant human bone morphogenetic protein-2/absorbable collagen sponge (rhBMP-2/ACS) on healing in 3-wall intrabony defects in dogs. *J Periodontol* 73(1), 63–72.

CRISAN, M., YAP, S., CASTEILLA, L., CHEN. C.W., CORSELLI, M., PARK, T.S., ANDRIOLO, G., SUN, B., ZHENG, B., ZHANG, L., NOROTTE, C., TENG, P.N., TRAAS, J., SCHUGAR, R., DEASY, B.M., BADYLAK, S., BUHRING, H.J., GIACOBINO, J.P., LAZZARI, L., HUARD, J., PÉAULT, B. (2008). A perivascular origin for mesenchymal stem cells in multiple human organs. *Cell Stem Cell* 3(3), 301–13.

CRIVELLATO, E., NICO, B., RIBATTI, D. (2007). Contribution of endothelial cells to organogenesis: A modern reappraisal of an old Aristotelian concept. *J Anat* 211(4), 415–27.

DISHER, D.E., JANMEY, P., WANG, Y.L. (2005). Tissue cells feel and respond to the stiffness of their substrate. *Science* 310(5751), 1139–43.

FOLKMAN, J., KLAGSBRUN, M., SASSE, J., WADZINKI, M., INGBER, D., VLODAVSKY I. (1988). A heparin binding angiogenic protein—basic fibroblast growth factor—is stored within basement membranes. *Am J Pathol* 130(2), 393–400.

FOPPIANO, S., HU, D., MARCUCIO, R.S. (2007). Signaling by bone morphogenetic proteins directs formation of an ectodermal signaling center that regulates craniofacial development. *Dev Biol* 312(1), 103–14.

FU, J., WANG, Y.-K., YANG, M.T., DESAI, R.A., YU, X., LIU, Z., CHEN, C.S. (2010). Mechanical regulation of cell function with geometrically modulated elastomeric substrates. *Nat Methods* 7, 733–36.

GIANNOBILE, W.V., RYAN, S., SHIH M.S., SU, D.L., KAPLAN, P.L., CHAN, T.C. (1998). Recombinant human osteogenic protein-1 (OP-1) stimulates periodontal wound healing in Class III furcation defects. *J Periodontol* 69(2), 129–37.

GOTTLOW, J., NYMAN, S., KARRING, T., LINDHE, J. (1984). New attachment formation as the result of controlled tissue regeneration. *J Clin Periodontol* 11(8), 494–503.

GRZESIK, W.J., KUZENTSOV, S.A., UZAWA, K., MANKANI, M., ROBEY, P.G., YAMAUCHI, M. (1998). Normal human cementum-derived cells: Isolation, clonal expansion, and in vitro and in vivo characterization. *J Bone Miner Res* 13(10), 1547–54.

HAKKI, S.S., FOSTER, B.L., NAGATOMO, K.J., BOZKURT, S.B., HAKKI, E.E., SOMERMAN, M.J., NOHUTCU, R.M. (2010). Bone morphogenetic protein-7 enhances cementoblast function in vitro. *J Periodontol* 81(11), 1663–74.

HELIOTIS, M., RIPAMONTI, U. (1994). Phenotypic modulation of endothelial cells by bone morphogenetic protein fraction in vitro. *In Vitro Cell Dev Biol Anim* 30A(6), 353–55.

HOGAN, B.L. (1996). Bone morphogenetic proteins in development. *Curr Opin Genet Dev* 6(4), 432–38.

HOWARD, P.S., MYERS, J.C., GORFIEN, S.F., MACARAK, E.J. (1991). Progressive modulation of endothelial phenotype during in vitro blood vessel formation. *Dev Biol* 146(2), 325–38.

JAFFE, E.A., NACHMAN, R.L., BECKER, C.G., MINICK, C.R. (1973). Culture of human endothelial cells derived from umbilical veins: Identification by morphologic and immunologic criteria. *J Clin Invest* 52(11), 2745–56.

KEITH, A. (1927). Concerning the origin and nature of osteoblasts. *Proc R Soc Med* 21(2), 301–8.

KHOURI, R.K., KOUDSI, B., REDDI, A.H. (1991). Tissue transformation into bone in vivo: A potential practical application. *JAMA* 266(14), 1953–55.

KLAR, R.M., DUARTE, R., DIX-PEEK, T., RIPAMONTI, U. (2014). The induction of bone formation by the recombinant human transforming growth factor-$\beta_3$. *Biomaterials* 35, 2773–88. doi: 10.1016/j.biomaterials.2013.12.062.

KOVACIC, J.C., BOEHM, M. (2009). Resident vascular progenitor cells: An emerging role for non-terminally differentiated vessel-resident cells in vascular biology. *Stem Cell Res* 2(1), 2–15.

LEVANDER, G. (1938). A study of bone regeneration. *Surg Gynec Obstet* 67(6), 705–14.

LIN, N.H., MENICANIN, D., MROZIK, K., GRONTHOS, S., BARTOLD, P.M. (2008). Putative stem cells in regenerative human periodontium. *J Periodont Res* 43(5), 514–23.

LIU, H.W., YACOBI, R., SAVION, N., NARAYANAN, A.S., PITARU, S. (1997). A collagenous cementum-derived attachment protein is a marker for progenitors of the mineralized tissue-forming cell lineage of the periodontal ligament. *J Bone Miner Res* 12(10), 1691–99.

LOZITO, T.P., TUAN, R.S. (2011). Mesenchymal stem cells inhibit both endogenous and exogenous MMPs via secreted TIMPs. *J Cell Physiol* 226(2), 385–96.

LUYTEN, F.P., CUNNINGHAM, N.S., MA, S., MUTHUKUMARAN, N., HAMMONDS, R.G., NEVINS, W.B., WOODS, W.I., REDDI, A.H. (1989). Purification and partial amino acid sequence of osteogenin, a protein initiating bone differentiation. *J Biol Chem* 264(23), 13377–80.

MAGAN, A., RIPAMONTI, U. (2013). Biological aspects of periodontal tissue regeneration: Cementogenesis and the induction of Sharpey's fibres. *South Afr Dent J* 68(7), 304–14.

MOEHL, T., RIPAMONTI, U. (1992). Primate dentine extracellular matrix induces bone differentiation in heterotopic sites of the baboon (*Papio ursinus*). *J Periodontal Res* 27(2), 92–96.

NAKAE, H., NARAYANAN A.S., RAINES, E., PAGE, R.C. (1991). Isolation and partial characterization of mitogenic factors from cementum. *Biochemistry* 30(29), 7047–52.

NARAYANAN, A.S., YONEMURA, K. (1993). Purification and characterization of a novel growth factor from cementum. *J Periodontal Res* 28(6 Pt 2), 563–65.

PARALKAR, V.M., NANDEDKAR, A.K.N., POINTER, R.H., KLEINMAN, H.K., REDDI, A.H. (1990). Interaction of osteogenin, a heparin binding bone morphogenetic protein, with type IV collagen. *J Biol Chem* 265(28), 17281–84.

PARALKAR, V.M., VUKICEVIC, S., REDDI A.H. (1991). Transforming growth factor β type 1 binds to collagen type IV of basement membrane matrix: Implications for development. *Dev Biol* 143(2), 303–8.

PITARU, S., SAVION, N., HEKMATI, H., OLSON, S., NARAYANAN, S.A. (1993). Molecular and cellular interactions of a cementum attachment protein with periodontal cells and cementum matrix components. *J Periodont Res* 28(6 Pt 2), 560–62.

RAMOSHEBI, L.N., RIPAMONTI, U. (2000). Osteogenic protein-1, a bone morphogenetic protein, induces angiogenesis in the chick chorioallantoic membrane and synergizes with basic fibroblast growth factor and transforming growth factor-beta1. *Anat Rec* 259(1), 97–107.

REDDI, A.H. (1981). Cell biology and biochemistry of endochondral bone development. *Coll Relat Res* 1(2), 209–26.

REDDI, A.H. (1984). Extracellular matrix and development. In K.A. Piez and A.H. Reddi (eds.), *Extracellular Matrix Biochemistry.* New York: Elsevier, pp. 375–412.

REDDI, A.H. (1997). Bone morphogenesis and modeling: Soluble signals sculpt osteosomes in the solid state. *Cell* 89(2), 159–61.

REDDI, A.H. (2000). Morphogenesis and tissue engineering of bone and cartilage: Inductive signals, stem cells, and biomimetic biomaterials. *Tissue Eng* 6(4), 351–59.

REDDI, A.H., HUGGINS, C. (1972). Biochemical sequences in the transformation of normal fibroblasts in adolescent rats. *Proc Natl Acad Sci USA* 69(6), 1601–5.

RIPAMONTI, U. (1988). Paleopathology in *Australopithecus africanus*: A suggested case of a 3-million-year-old prepubertal periodontitis. *Am J Phys Anthropol* 76(2), 197–210.

RIPAMONTI, U. (1989). The hard evidence of alveolar bone loss in early hominids of Southern Africa: A short communication. *J Periodontol* 60(2), 118–20.

RIPAMONTI, U. (1991). Bone induction in non-human primates: An experimental study on the baboon. *Clin Orthop* 269, 284–94.

RIPAMONTI, U. (2003). Osteogenic proteins of the transforming growth factor-β superfamily. In H.L. Henry and A.W. Norman (eds.), *Encyclopedia of Hormones*. San Diego: Academic Press, pp. 80–86.

RIPAMONTI, U. (2005). Bone induction by recombinant human osteogenic protein-1 (hOP-1, BMP-7) in the primate *Papio ursinus* with expression of mRNA of gene products of the TGF-β superfamily. *J Cell Mol Med* 9(4), 911–28.

RIPAMONTI, U. (2006). Soluble osteogenic molecular signals and the induction of bone formation. *Biomaterials* 27(6), 807–822.

RIPAMONTI, U. (2007). Recapitulating development: A template for periodontal tissue engineering. *Tissue Eng* 13(1), 51–71.

RIPAMONTI, U. (2009). Biomimetism, biomimetic matrices and the induction of bone formation. *J Cell Mol Med* 13(9B), 2953–72.

RIPAMONTI, U., CROOKS, J., MATSABA, T., TASKER, J. (2000b). Induction of endochondral bone formation by recombinant human transforming growth factor-$\beta_2$ in the baboon (*Papio ursinus*). *Growth Factors* 17(4), 269–85.

RIPAMONTI, U., CROOKS, J., TEARE, J., PETIT, J.-C., RUEGER, D.C. (2002). Periodontal tissue regeneration by human recombinant osteogenic protein-1 in periodontally-induced furcation defects of the primate *Papio ursinus*. *South Afr J Sci* 98(7–8), 361–68.

RIPAMONTI, U., DUARTE R., FERRETTI, C. (2014). Re-evaluating the induction of bone formation in primates. *Biomaterials* 35, 9407–22.

RIPAMONTI, U., DUNEAS, N., VAN DEN HEEVER, B., BOSCH, C., CROOKS, J. (1997). Recombinant transforming growth factor-$\beta_1$ induces endochondral bone in the baboon and synergizes with recombinant osteogenic protein-1 (bone morphogenetic protein-7) to initiate rapid bone formation. *J Bone Miner Res* 12(10), 1584–95.

RIPAMONTI, U., FERRETTI, C., HELIOTIS, M. (2006). Soluble and insoluble signals and the induction of bone formation: Molecular therapeutics recapitulating development. *J Anat* 209(6), 447–68.

RIPAMONTI, U., FERRETTI, C., TEARE, J., BLANN, L. (2009c). Transforming growth factor-β isoforms and the induction of bone formation: Implications for reconstructive craniofacial surgery. *J Craniofac Surg* 20(5), 1544–55.

RIPAMONTI, U., HELIOTIS, M., FERRETTI, C. (2007). Bone morphogenetic proteins and the induction of bone formation: From laboratory to patients. *Oral Maxillofac Surg Clin North Am* 19(4), 575–89, vii.

RIPAMONTI, U., HELIOTIS, M., RUEGER, D.C., SAMPATH, T.K. (1996). Induction of cementogenesis by recombinant human osteogenic protein-1 (hOP-1/BMP-7) in the baboon (*Papio ursinus*). *Arch Oral Biol* 41(1), 121–26.

RIPAMONTI, U., HELIOTIS, M., VAN DEN HEEVER, B., REDDI, A.H. (1994). Bone morphogenetic proteins induce periodontal regeneration in the baboon (*Papio ursinus*). *J Periodont Res* 29(6), 439–45.

RIPAMONTI, U., HERBST, N.N., RAMOSHEBI, L.N. (2005). Bone morphogenetic proteins in craniofacial and periodontal tissue engineering: Experimental studies in the non-human primate *Papio ursinus*. *Cytokine Growth Factor Rev* 16(3), 357–68.

RIPAMONTI, U., MA, S., CUNNINGHAM, N.S., YEATES, L., REDDI, A.H. (1992). Initiation of bone regeneration in adult baboons by osteogenin, a bone morphogenic protein. *Matrix* 12(5), 369–80.

RIPAMONTI, U., PARAK, R., PETIT, J.C. (2009b). Induction of cementogenesis and periodontal ligament regeneration by recombinant human transforming growth factor-$\beta_3$ in Matrigel with rectus abdominis responding cells. *J Periodont Res* 44(1), 81–87.

RIPAMONTI, U., PETIT, J.-C. (2009). Bone morphogenetic proteins, cementogenesis, myoblastic stem cells and the induction of periodontal tissue regeneration. *Cytokine Growth Factor Rev* 20(5–6), 489–99.

RIPAMONTI, U., PETIT, J.-C., GROSSMAN, E.S. (1989). The hard periodontal tissues of the Australopithecinae. *Proc Electron Microsc Soc South Afr* 19, 141–42.

RIPAMONTI, U., PETIT, J.C., TEARE, J. (2009a). Cementogenesis and the induction of periodontal tissue regeneration by osteogenic proteins of the transforming growth factor-$\beta$ superfamily. *J Periodont Res* 44(2), 141–52.

RIPAMONTI, U., RAMOSHEBI, L.N., CROOKS, J., PETIT, J.-C., RUEGER D.C. (2001). Periodontal tissue regeneration by combined applications of recombinant human osteogenic protein-1 and bone morphogenetic protein-2: A pilot study in Chacma baboons (*Papio ursinus*). *Eur J Oral Sci* 109(4), 241–48.

RIPAMONTI, U., RAMOSHEBI, L., PATTON, J., MATSABA, T., TEARE, J., RENTON, L. (2004). Soluble signals and insoluble substrata: Novel molecular cues instructing the induction of bone. In E.J. Massaro and J.M. Rogers (eds.), *The Skeleton: Biochemical, Genetic, and Molecular Interactions in Development and Homeostasis*. Totowa, NJ: Humana Press, pp. 217–27.

RIPAMONTI, U., RAMOSHEBI, L.N., TEARE, J., RENTON, L., FERRETTI, C. (2008). The induction of endochondral bone formation by transforming growth factor-$\beta_3$: Experimental studies in the non-human primate *Papio ursinus*. *J Cell Mol Med* 12(3), 1029–48.

RIPAMONTI, U., REDDI, A.H. (1994). Periodontal regeneration: Potential role of bone morphogenetic proteins. *J Periodont Res* 29(4), 225–35.

RIPAMONTI, U., REDDI, A.H. (1995). Bone morphogenetic proteins: Applications in plastic and reconstructive surgery. *Adv Plast Reconstr Surg* 11, 47–65.

RIPAMONTI, U., REDDI, A.H. (1997). Tissue engineering, morphogenesis, and regeneration of the periodontal tissues by bone morphogenetic proteins. *Crit Rev Oral Biol Med* 8(2), 154–63.

RIPAMONTI, U., RENTON, L. (2006). Bone morphogenetic proteins and the induction of periodontal tissue regeneration. *Periodontol 2000* 41, 73–87.

RIPAMONTI, U., RODEN, L.C. (2010). Induction of bone formation by transforming growth factor-$\beta_2$ in the non-human primate *Papio ursinus* and its modulation by skeletal muscle responding cells. *Cell Prolif* 43(3), 207–18.

RIPAMONTI, U., VAN DEN HEEVER, B., CROOKS, J., TUCKER, M.M., SAMPATH, T.K., RUEGER, D.C., REDDI, A.H. (2000a). Long-term evaluation of bone formation by osteogenic protein-1 in the baboon and relative efficacy of bone-derived bone morphogenetic proteins delivered by irradiated xenogeneic collagenous matrices. *J Bone Miner Res* 15(9), 1798–809.

SAMPATH, T.K., REDDI, A.H. (1981). Dissociative extraction and reconstitution of extracellular matrix components involved in local bone differentiation. *Proc Natl Acad Sci USA* 78(12), 7599–603.

SAMPATH, T.K., REDDI, A.H. (1983). Homology of bone-inductive proteins from human, monkey, bovine, and rat extracellular matrix. *Proc Natl Acad Sci USA* 80(21), 6591–95.

SCHNITZLER, C.M., RIPAMONTI, U., MESQUITA, J.M. (1993). Histomorphometry of iliac crest trabecular bone in adult male baboons in captivity. *Calcif Tissue Int* 52(6), 447–54.

SIGURDSSON, T.J., LEE, M.B., KUBOTA, K., TUREK, T.J., WOZNEY, J.M., WIKESJÖ, U.M. (1995a). Periodontal repair in dogs: Recombinant human bone morphogenetic protein-2 significantly enhances periodontal regeneration. *J Periodontol* 66(2), 131–38.

SIGURDSSON, T.J. TATAKIS, D.N., LEE, M.B., WIKESJÖ, U.M. (1995b). Periodontal regenerative potential of space-providing expanded polytetrafluoroethylene membranes and recombinant human bone morphogenetic proteins. *J Periodontol* 66(6), 511–21.

TAKAHASHI K., TANABE, K., OHNUKI, M., NARITA, M., ICHISAKA, T., TOMODA, K., YAMANAKA, S. (2007). Induction of pluripotent stem cells from adult human fibroblasts by defined factors. *Cell* 131(5), 861–72.

TAKAHASHI, K., YAMANAKA, S. (2006). Induction of pluripotent stem cells from mouse embryonic and adult fibroblast cultures by defined factors. *Cell* 126(4), 663–76.

TEARE, J.A., RAMOSHEBI, L.N., RIPAMONTI, U. (2008). Periodontal regeneration by recombinant human transforming growth factor-$\beta_3$ in *Papio ursinus*. *J Perodont Res* 43(1), 1–8.

THESLEFF, I. (2006). The genetic basis of tooth development and dental defects. *Am J Med Genet A* 140(23), 2530–35.

THESLEFF, I., NIEMINEN, P. (1996). Tooth morphogenesis and cell differentiation. *Curr Opin Cell Biol* 8(6), 844–50.

THESLEFF, I., SHARPE, P. (1997). Signalling networks regulating dental development. *Mech Dev* 67(2), 111–23.

THOMADAKIS, G., RAMOSHEBI L.N., CROOKS, J., RUEGER, D.C., RIPAMONTI, U. (1999). Immunolocalization of bone morphogenetic protein-2 and -3 and osteogenic protein-1 during murine tooth root morphogenesis and in other craniofacial structures. *Eur J Oral Sci* 107(5), 368–77.

TRUETA, J. (1963). The role of the vessels in osteogenesis. *J Bone Joint Surg* 45B, 402–18.

TURING, A.M. (1952). The chemical basis of morphogenesis. *Philos Trans R Soc Lond* 237, 37–41.

URIST, M.R. (1965). Bone: Formation by autoinduction. *Science* 150(3698), 893–99.

URIST, M.R, DOWELL, T.A., HAY, P.H., STRATES, B.S. (1968). Inductive substrates for bone formation. *Clin Orthop Relat Res* 59, 59–96.

URIST, M.R., SILVERMAN, B.F., BURING, K., DUBUC, F.L., ROSENBERG, J.M. (1967). The bone induction principle. *Clin Orthop Relat Res* 53, 243–83.

USAS, A., MAČIULAITIS, J., MAČIULAITIS, R., JAKUBONIENĖ, N., MILAŠIUS, A., HUARD, J. (2011). Skeletal muscle-derived stem cells: Implications for cell-mediated therapies. *Medicina* 47(9), 469–79.

VAINIO, S., KARAVANOVA, I., JOWETT, A., THESLEFF, I. (1993). Identification of BMP-4 as a signal mediating secondary induction between epithelial and mesenchymal tissues during early tooth development. *Cell* 75(1), 45–58.

VLODAVSKY, I., FOLKMAN, J., SULLIVAN, R., FRIDMAN, R., ISHAI-MICHAELI, R., SASSE, J., KLAGSBRUN, M. (1987). Endothelial cell-derived basic fibroblast growth factor: Synthesis and deposition into subendothelial extracellular matrix. *Proc Natl Acad Sci USA* 84(8), 2292–96.

WU, D., IKEZAWA, K., PARKER, T., SAITO, M., NARAYANAN, A.S. (1996). Characterization of a collagenous cementum-derived attachment protein. *J Bone Miner Res* 11(5), 686–92.

YONEMURA, K., NARAYANAN, A.S., MIKI, Y., PAGE, R.C., OKADA, H. (1992). Isolation and partial characterization of a growth factor from cementum. *Bone Miner* 18(3), 187–98.

ZHAO, M., BERRY, J.E., SOMERMAN M.J. (2003). Bone morphogenetic protein-2 inhibits differentiation and mineralization of cementoblasts in vitro. *J Dent Res* 82(1), 23–27.

ZHENG, B., CAO, B., CRISAN, M., SUN, B., LI, G., LOGAR, A., YAP, S., POLLETT, J.B., DROWLEY, L., CASSINO, T., GHARAIBEH, B., DEASY, B.M., HUARD, J., PÉAULT, B. (2007). Prospective identification of myogenic endothelial cells in human skeletal muscle. *Nat Biotechnol* 25(9), 1025–34.

# Index

T - #0432 - 071024 - C212 - 234/156/10 - PB - 9780367377403 - Gloss Lamination